The Challenge of
CMC Regulatory Compliance
for Biopharmaceuticals

The Challenge of
CMC Regulatory Compliance
for Biopharmaceuticals

John Geigert

BioPharmaceutical Quality Solutions
Carlsbad, California

Kluwer Academic / Plenum Publishers
New York, Boston, Dordrecht, London, Moscow

Library of Congress Cataloging-in-Publication Data

Geigert, J. (John), 1948–
    The challenge of CMC regulatory compliance for biopharmaceuticals/John Giegert.
        p.   cm.
    Includes bibliographical references and index.
    ISBN 0-306-48040-9
        1. Pharmaceutical biotechnology.   2. Pharmaceutical biotechnology industry—Law and
    legislation.   3. Pharmaceutical biotechnology—Quality control.   I. Title.

    RS380.G45 2004
    615′.19—dc22

                                                                        2003061969

ISBN 0-306-48040-9

©2004 Kluwer Academic / Plenum Publishers, New York
233 Spring Street, New York, New York 10013

http://www.kluweronline.com

10 9 8 7 6 5 4 3 2 1

A C.I.P. record for this book is available from the Library of Congress

Permissions for books published in Europe: *permissions@wkap.nl*
Permissions for books published in the United States of America: *permissions@wkap.com*

Printed in the United States of America

Preface

My primary purpose for writing this book was much more than to provide another information source on Chemistry, Manufacturing & Controls (CMC) that would rapidly become out of date. My primary purpose was to provide insight and practical suggestions into a common sense business approach to manage the CMC regulatory compliance requirements for biopharmaceuticals. Such a common sense business approach would need (1) to be applicable for all types of biopharmaceutical products both present and future, (2) to address the needs of a biopharmaceutical manufacturer from the beginning to the end of the clinical development stages and including post-market approval, and (3) to be adaptable to the constantly changing CMC regulatory compliance requirements and guidance. Trying to accomplish this task was a humbling experience for this author!

In Chapter 1, the CMC regulatory process is explained, the breadth of products included under the umbrella of biopharmaceuticals are identified, and the track record for the pharmaceutical and biopharmaceutical industry in meeting CMC regulatory compliance is discussed. In Chapter 2, while there are many CMC commonalities between biopharmaceuticals and chemically-synthesized pharmaceuticals, the significant differences in the way the regulatory agencies handle them are examined and the reasons for why such differences are necessary is discussed. Also, the importance of CMC specific meetings, as well as an ongoing CMC dialogue, with the FDA is stressed. In Chapter 3, the three key elements for a successful, corporate CMC regulatory compliance strategy are introduced, along with an explanation of the concept of the minimum CMC continuum. In Chapter 4, the importance of compliance with current good manufacturing practices, starting with the first human clinical studies in Phase 1, is discussed. The 'biopharm' in the manufacturing and control of biopharmaceuticals is primarily in the manufacture of the active pharmaceutical ingredient (the API, also referred to as

v

biological substance), which is the reason the CMC regulatory compliance is thoroughly discussed for the recombinant source material in Chapter 5, the production systems in Chapter 6, and the purification processes in Chapter 7. The downstream drug product manufacturing (typically formulation of the API and filling into containers for patient application) is fairly similar whether the product is a chemically-synthesized drug or a biopharmaceutical, which is why in Chapter 8 only those significant CMC regulatory compliance issues that differ between the product types are discussed. In Chapter 9, the important aspects of product characterization, release tests and specifications, and stability profiles and expiration dating, are thoroughly examined.   Manufacturing processes are dynamic, but all changes need to be controlled and, most importantly for a biopharmaceutical, the product manufactured after the process changes needs to be comparable to the product manufactured before the change, as outlined in Chapter 10. In Chapter 11, because of the heavy reliance on outsourcing with biopharmaceuticals, how to properly manage the CMC regulatory compliance issues is examined. In Chapter 12, I attempt to 'crystal ball' some future CMC regulatory compliance issues for biopharmaceuticals.

While the focus of this book is to address the concerns identified by the U.S, Food and Drug Administration (FDA), I have incorporated many of the CMC regulatory compliance requirements and guidance documents provided by the European Agency for the Evaluation of Medicinal Products (EMEA) and the International Conference on Harmonization (ICH), where appropriate.

To put to rest the statement that 'the FDA doesn't provide enough CMC regulatory compliance for biopharmaceuticals', I have attempted to demonstrate the abundance of such guidance. Nearly all of the 200 plus references listed in the back of this book are from the U.S. FDA. I have attempted to quote heavily from these resources. But this abundance of CMC regulatory compliance information for biopharmaceuticals has both a good news and a bad news aspect.

The 'good news' is that CMC regulatory compliance no longer has to be a 'mystery', understood by only a few people in the regulatory affairs departments of our companies. Less than 10 years ago, I can remember how difficult and time consuming it was to obtain copies of the necessary FDA documents, let alone any regional or international ones. The biopharmaceutical industry owes much thanks to the regulatory agencies for their foresight and commitment in getting the information into the hands of those who need it. Through means of the internet, anybody can now download these documents for review from anywhere and at any time. It is for this reason that I have provided website addresses for the regulations and guidance documents that were used in the preparation of this book.

The 'bad news' is that there can be too much CMC regulatory compliance information available, 'an information overload'. At times assistance is needed to work through all of the help and guidance publicly available, especially in evaluating as to whether it has any useful application to our biopharmaceutical product at its current stage of drug development. This is where the help of an experience consultant in CMC regulatory compliance becomes invaluable. Another aspect of the 'bad news' is that there continues to be much change in this area of CMC regulatory compliance for biopharmaceuticals. If we are not careful our information can rapidly become dated. It is for this reason that I have provided websites in Chapter 12 that will assist you in obtaining ongoing CMC regulatory compliance updates.

It is my desire that the common sense business approach to CMC regulatory compliance presented in this book will be of help to biopharmaceutical companies today and for many years to come.

ACKNOWLEDGEMENTS

Many people over my 25 years of service in the biopharmaceutical industry have impacted my understanding and have indirectly contributed to the writing of this book. I would like to especially acknowledge my friends and colleagues at my former companies – Cetus (now Chiron), Immunex (now Amgen) and IDEC Pharmaceuticals (soon to be Biogen Idec) – for the insights and experiences that we shared. I would also like to acknowledge my new friends and colleagues in the many biopharmaceutical companies that I now serve as a consultant – for the many CMC regulatory compliance challenges that we wrestle with.

A special expression of appreciation goes to my wife, Nicki, who understood the time commitment and mental exhaustion that comes along with trying to write a book of this magnitude, and for her patient support and encouragement through this entire process.

John Geigert, Ph.D., RAC
President
BioPharmaceutical Quality Solutions
Carlsbad, CA 92009

Contents

Chapter 1
Biopharmaceutical CMC Regulatory Compliance: What is It?

Chapter 2
Are Biopharmaceuticals Really Different?

Chapter 3
Developing the Corporate CMC Regulatory Compliance Strategy

Chapter 4
Can't Ignore cGMPs

Chapter 5
Recombinant Source Material: Master/Working Banks

Chapter 6
Production: Expansion of the Recombinant Organism and Expression of the Biopharmaceutical

Chapter 7
Purification of the Biopharmaceutical

Chapter 8
Biopharmaceutical Drug Product Manufacturing

Chapter 9
Physicochemical/Biological Analysis of the Biopharmaceutical Product

Chapter 10
Managing Process Changes – Demonstrating Product Comparability

Chapter 11
Biopharmaceutical CMC Outsourcing

Chapter 12
Concluding Thoughts on Biopharmaceutical CMC Regulatory Compliance

TABLES

FIGURES

Biopharmaceutical CMC Regulatory Compliance: What is It?

"We expect the first part of the 21st century to usher in a new golden age of pharmaceuticals. It will begin with the introduction of a powerful arsenal of weapons against the 200 or so diseases we call cancer. We don't predict that cancer will be conquered. However, we do believe that between new treatments and drugs that modify the eating, drinking and smoking behaviours that lead to cancer, rates of the most common forms of this disease will plummet."

Popular Mechanics prediction of some miracles by the year 2050, published in their January 2000 issue

1. DEFINING OUR TERMS

Our growing scientific knowledge and technical experiences have given rise to complex biopharmaceutical manufacturing processes and products. This increase in complexity has led to an ever-growing and evolving set of regulatory agency expectations regarding CMC activities. It is most critical that companies understand what defines compliance for their type of biopharmaceutical product.

Three major terms need to be defined: 'biopharmaceutical', 'CMC' and 'CMC regulatory compliance'.

1.1. What is a 'Biopharmaceutical'?

It wasn't too long ago, actually less than 25 years, that the term 'biopharmaceutical' was not part of anybody's common vocabulary. Regulatory documents issued in the 1980's used terms such as 'recombinant DNA-derived products' and 'biotechnology-derived products' (or BDP).[1, 2, 3] The term 'biopharmaceutical' did not begin appearing in regulatory documents until the later 1990's:

> "Biotechnology-derived pharmaceuticals (biopharmaceuticals) were initially developed in the early 1980's."[4]

Today, everyone is using the term. We have biopharmaceutical companies, biopharmaceutical consultants (like myself), and of course many drug products that are referred to as biopharmaceuticals.

So what exactly is a 'biopharmaceutical' and how is it different from other drug products? Drugs, a term that can be applied in general to all therapeutic products, typically refers to pharmaceuticals derived either from chemical synthesis (e.g., Zocor®, Prozac®, aspirin), or from 'classical' fermentation (e.g., antibiotics such as penicillin and gentamicin), or from natural sources (e.g., taxol from Yew trees). Biologics are those therapeutic products that are derived from living sources and include vaccines, blood and blood products, allergenic extracts and tissues. Biopharmaceuticals are those drugs and biologics that are created through the genetic manipulation of foreign DNA into living organisms. A biopharmaceutical is therefore a biotechnology-derived drug. Genetic engineering of the living host used to produce the drug is mandatory.

Early on there were the obvious concerns about this new technology of genetic engineering, and even a brief moratorium on developing genetically engineered organisms was observed in the late 1970's, but through the coordinated efforts by the governmental agencies and the pharmaceutical industry these concerns were readily addressed. Today, applying genetic engineering to produce new medicinal products is common in the pharmaceutical industry. The value of these biotechnology-derived drug products to improve patient health is widely accepted by the public. New treatments for rheumatoid arthritis, multiple sclerosis and a variety of cancers have been the positive impact of the biopharmaceutical products on the market today.

1.2. What is 'CMC'

Chemistry, Manufacturing and Controls (CMC) is the body of information that defines:

- the manufacturing facility and all of its support utilities (their design, qualification, operation, maintenance)

- the process equipment and materials used in manufacturing (design, qualification, validation, operation, maintenance)

- the manufacturing process itself (definition, validation, consistency)

- the personnel involved in manufacturing and quality (adequate numbers and competency)

- the chemistry of the product (characterization and proof of structure)

- Quality Control release testing, specifications and stability of the product

- Quality Assurance release and rejection of materials and product

- All of the controls, documentation, and training necessary to ensure that all of the above is properly and effectively carried out

CMC includes the chemistry, and the manufacturing, and the controls for the biopharmaceutical.

1.3. What is 'CMC Regulatory Compliance'?

'CMC regulatory compliance' is ensuring that all of the above CMC practices are carried out in agreement with regulatory agencies requirements and expectations. Since such requirements and expectations change with time, CMC regulatory compliance is to ensure that all of the above CMC practices are updated accordingly.

In addition, CMC regulatory compliance is to ensure that if the company has made any specific CMC commitment to the regulatory agencies, either verbally or in writing, that such CMC practices are carried out.

2. UNDER THE BIOPHARMACEUTICAL UMBRELLA

Biopharmaceuticals, therapeutic products created through the genetic manipulation of living things, include recombinant DNA-derived proteins and monoclonal antibodies, along with gene therapies, and now bioengineered animals and plants ("transgenics").

2.1. Recombinant DNA-Derived Proteins

The recombinant proteins, also known as rDNA-derived proteins, represent the largest group of biopharmaceuticals approved to date by the FDA. Table 1 presents just some of the many therapeutic products in this group.

Many of these biopharmaceuticals have replaced their biologic equivalent, and for good reasons. For example, Factor VIII which is used in treatment of blood clotting in hemophiliacs was previously isolated from human blood. Not only was there supply issues (about 10 tons of processed plasma yielded only about 10 mg of Factor VIII), but also there were past safety concerns (blood pools contaminated with the virus for AIDS due to insufficient donor controls). Recombinant human Factor VIII is now produced by the hundreds of grams in bioreactors using a genetically engineered cell lines that are absent of human viruses. Another example is human growth hormone which was previously isolated from the pituitary glands from human cadavers. Not only was there limited supply (a few grams of hormone extracted from 100 human glands), but also there were major, difficult to test for, safety concerns (the whole issue of human viruses present in the tissues and more specifically the possible presence of the neurological disease, Creutzfeld-Jakob Disease, CJD). Recombinant human growth hormone is now produced by the kilograms in bioreactors using genetically engineered cell lines that are absent of human viruses and CJD.

Some of these recombinant DNA-derived proteins are improvements over the properties of the natural biologic equivalent. For example, through genetic engineering of the natural gene itself, product stability has been enhanced (e.g., substitution of the cysteine codon with a serine codon to prevent disulfide scrambling in recombinant IL-2) and improved bioavailability has been obtained (e.g., Human Insulin Lispro which has two codons reversed compared Human Insulin).

Table 1. Some FDA approved recombinant DNA-derived human protein biopharmaceutical products; listed on their websites: www.fda.gov/cber/products.htm, www.fda.gov/cder/approval/index.htm

Biopharmaceutical	Manufacturing and Product Information
Human Insulin Humulin®	Recombinant human insulin; 51 amino acid protein of molecular weight of 5808 daltons; recombinant *Escherichia coli* (r*E. coli*)
Human Insulin Lispro Humalog®	Two codons reversed in human insulin gene coding for amino acid positions 28 and 29; same molecular weight as human insulin; r*E. coli*
Human Interleukin 2 Proleukin®	Recombinant human interleukin-2 with a serine genetically substituted for cysteine at amino acid position 125; molecular weight of 15.5 kilodalton; r*E. coli*
Anakinra Kineret®	Recombinant human interleukin-1 receptor antagonist; 153 amino acids of molecular weight of 17.3 kilodaltons; r*E. coli*
Pegfilgrastim Neulasta®	Recombinant human granulocyte colony stimulating factor (G-CSF) chemically conjugated to polyethylene glycol; r*E. coli*
Denileukin Diftitox ONTAK®	Recombinant fusion protein consisting of human interleukin 2 and diphtheria toxin fragments A and B; approximate molecular weight of 58 kilodaltons; r*E. coli*
Hepatitis B Vaccine Recombivax HB®	Recombinant hepatitis B virus surface antigen; recombinant *Saccharomyces cerevisiae*
Sargramostim Leukine®	Recombinant human granulocyte-macrophage colony stimulating factor (GM-CSF); 3 molecular species of approximate molecular weight 19.5, 16.8 and 15.5 kilodaltons; r*S. cerevisiae*

Alefacept Amevive®	Recombinant dimeric fusion protein consisting of the extracellular CD2-binding portion of the human leukocyte function antigen-3 (LFA-3) linked to the Fc portion of human IgG1; 91.4 kilodaltons; recombinant Chinese hamster ovary (rCHO)
Etanercept Enbrel®	Recombinant dimeric fusion protein consisting of the human TNF receptor and the Fc portion of human IgG1; 934 amino acids in the protein of approximate molecular weight of 150 kilodaltons; rCHO
Antihemophilic Factor ReFacto®	Recombinant human Factor VIII; 1438 amino acid protein of approximate molecular weight of 170 kilodaltons; rCHO
Tenecteplase TNKase®	Six codons changed on human tissue plasminogen activator (tPA) gene; 527 amino acid protein; rCHO cells
Darbepoetin Alfa Aranesp®	Two codons changed on human EPO gene yielding two amino acids for additional N-linked oligosaccharide chains; 165-amino acid protein of approximate molecular weight of 37 kilodaltons; rCHO
Coagulation Factor VIIa NovoSeven®	Recombinant human Factor VIIa; 406 amino acid protein of approximate molecular weight of 50 kilodaltons; recombinant baby hamster kidney (rBHK); converted to active two-chain form during a chromatographic process step
Drotrecogin Alfa Xigris®	Recombinant human activated Protein C; approximate molecular weight of 55 kilodaltons; recombinant HEK293 human cell line

2.2. Monoclonal Antibodies

The monoclonal antibodies, also known as clonal DNA-derived immunoglobulins, represent another large group of biopharmaceuticals approved to date by the FDA. Table 2 presents just some of the many therapeutic products in this group.

The first therapeutic monoclonal antibody approved by the FDA was OKT-3® in 1986, which was manufactured using hybridoma cells (cells prepared by fusing mouse antibody-producing cells with mouse myeloma cells to impart immortality) and mouse colonies. Purification of the collected ascitic fluid yielded the monoclonal antibody having fully murine amino acid sequences. But both the method of manufacture (the difficulty and challenges of using mouse colonies to produce an adequate supply) and the resulting product (a murine monoclonal antibody that could trigger undesired immunological responses when injected), initially limited the utility of this group of biopharmaceuticals.

However, the ability to detect and isolate the heavy and light chain genes for human antibodies coupled with the advances in carrying out genetic engineering on the genes or their fragments, opened the door first for chimeric (mixture of murine and human amino acid sequences), then humanized (minor murine sequences, mostly human amino acid sequences), and now fully human monoclonal antibodies to be considered, and provided a way to minimize immunogencity and increase serum lifetimes. Taking these genes and transfecting them into the chromosome of a production cell line adapted for growth in a stainless steel bioreactor (e.g., Chinese hamster ovary cells) changed the manufacturing strategy for these products and opened up their availability, allowing these monoclonal antibodies now to be manufactured in kilogram amounts.

Table 2. Some FDA approved monoclonal antibody biopharmaceuticals; listed on their websites: www.fda.gov/cber/products.htm, www.fda.gov/cder/approval/index.htm

Biopharmaceutical	Manufacturing and Product Information
Muromonmab-CD3 Orthoclone OKT-3®	Murine monoclonal antibody with affinity for CD-3 cell surface protein ; manufactured from a murine hybridoma and purified from mouse ascites
Rituximab Rituxan®	Chimeric (mouse/human) IgG1 monoclonal antibody with affinity for CD20 cell surface protein overexpressed in malignant B cells; recombinant Chinese hamster ovary (rCHO)
Infliximab Remicade®	Chimeric IgG1 monoclonal antibody with affinity for human turmor necrosis factor alpha; manufactured using a recombinant mouse myeloma cell line (rSP2/0) by continuous perfusion
Basiliximab Simulect®	Chimeric IgG1 monoclonal antibody with affinity for interleukin-2 receptor; recombinant mouse myeloma cells
Daclizumab Zenapax®	Humanized IgG1 monoclonal antibody with affinity for interleukin-2 receptor; recombinant cell line
Trastuzumab Herceptin®	Humanized IgG1 monoclonal antibody with affinity for HER2 cell surface protein overexpressed in breast cancer cells; rCHO
Palivizumab Synagis®	Humanized IgG1 monoclonal antibody with affinity for respiratory syncytial virus (RSV); recombinant mouse myeloma cells (rNS0)
Alemtuzumab Campath®	Humanized IgG1 monoclonal antibody with affinity for CD52 cell surface protein expressed on leukemic cells; rCHO
Gemtuzumab Ozogamicin Mylotarg®	Humanized IgG4 monoclonal antibody linked to the cytotoxic antitumor antibiotic, calicheamicin; with affinity for CD33 cell surface protein overexpressed on leukemic cells; rNS0
Ibritumomab Tiuxetan Zevalin®	Murine IgG1 monoclonal antibody linked to yttrium-90; with affinity for CD20 cell surface protein overexpressed in malignant B cells; rCHO
Tositumomab Bexxar®	Murine IgG2a monoclonal antibody linked to iodine-131; with an affinity for CD20 cell surface protein overexpressed in malignant B cells; mammalian cells
Adalimumab Humira®	Human IgG1 monoclonal antibody with affinity for human tumor necrosis factor; manufacture in recombinant mammalian cells

2.3. Gene Therapy

Gene therapy is a medical intervention based on modification of the genetic material of living human cells and its subsequent expression in vivo. Gene therapy can be defined by the type of vector used (the vector can be of any variety of viral types or can be non-viral, such as a plasmid), or by the product gene components (the therapeutic gene or genetic material contained within the vector can be protein coding, protein non-coding, or regulatory elements such as nucleic acid sequences that function to enhance or promote the expression of the therapeutic gene), or by mode of gene therapy product administration (in vivo or ex vivo products depending upon whether they are directly administered to the patient or introduced outside the patient, respectively). The ex vivo gene therapy can be further defined as autologous (the patient's own cells) or allogeneic (cells from another donor or cell line). It has been over 10 years since the first gene therapy clinical trials were started, and there are currently over 600 hundred gene therapy clinical studies in over 20 countries involving more than 3000 patients.

Vectors that are involved in gene transfer (i.e., the viral and plasmid vectors) are gene therapy biopharmaceuticals. A few examples are provided in Table 3; however, to date, none of these gene therapy biopharmaceuticals have received FDA approval for marketing.

Table 3. Examples of some gene therapy clinical studies; listed on the Journal of Gene Medicine website: www.wiley.co.uk/genetherapy/clinical

Genes Involved in Gene Therapy	% of Clinical Protocols (Top 5)
Cytokine	~ 21%
Multiple Genes	~16%
Antigen	~12%
Deficiency	~11%
Tumor Suppressor	~ 9%

Vectors Used in Gene Therapy	% of Clinical Protocols (Top 5)
Retrovirus	~34%
Adenovirus	~27%
Lipofection	~ 12%
Naked/Plasmid DNA	~ 11%
Pox Virus	~ 6%

2.4. Animal/Plant Transgenics

Biopharmaceuticals derived from bioengineered animals or plants (transgenics, the altering of animal or plant chromosomal DNA so that the new genetically modified organism contains a gene from another organism) are also moving through the drug development process. Transgenic animals such as goats, pigs, cows and rabbits produce biopharmaceuticals in their milk; while transgenic chickens express the biopharmaceutical in the eggs they lay. Transgenic plants such as corn and tobacco produce biopharmaceuticals in the kernel or leaf, respectively. Commercial interest in this technology is due to these production systems having the potential capacity to produce very large quantities of products at possibly lower production costs. For bioengineered plants, there is also the advantage reduced risks compared with mammalian cell systems (e.g., viruses and transmissible spongiform encephalopathies).

However, to date, biopharmaceuticals from neither bioengineered animals nor plants have received FDA approval for marketing.

2.5. Rapid Pace of Biopharmaceutical Development

Biopharmaceuticals have amassed an impressive market value in only two decades. Seven biopharmaceuticals are now included among the 50 top-selling marketed therapeutic products, with each product near or exceeding $1 billion in annual worldwide sales:

Biopharmaceutical	FDA Approved Use
Recombinant erythropoietin	red blood cell proliferation
Recombinant interferon alpha	treatment for chronic hepatitis C
Recombinant G colony stimulating factor	white blood cell proliferation
Recombinant human insulin	diabetes
Recombinant human growth hormone	growth factor
Monoclonal antibody to CD20 receptor	non-Hodgkin's lymphoma
Recombinant hepatitis B vaccine	vaccination against hepatitis B virus

One shouldn't underestimate the rapid pace of biopharmaceutical development. Two decades ago, monoclonal antibodies were considered to be 'magic bullets', administered in only microgram amounts. Today, monoclonal antibodies are widely used therapeutic products administered at up to gram quantities. Who can tell where gene therapy or transgenics will be in two decades from now? And who can tell of tomorrow's discoveries and the challenges they will place on the FDA and other regulatory agencies to appropriately regulate (e.g., the development in nanotechnology, in which multiple predetermined and controlled delivery of biopharmaceuticals can be placed on a chip).

3. REGULATORY DEVELOPMENT OF BIOPHARMACEUTICALS

Compared to the history of drugs and biologics, biopharmaceuticals are still relatively new in the marketplace. Both aspirin and smallpox vaccines were already available by the early 1900's. However, the first biopharmaceuticals did not enter human clinical studies until the 1970's, and then the first FDA approved biopharmaceutical (which was recombinant human insulin) did not occur until the early 1980's.

3.1. The Drug Development Process

Whether a drug, a biologic or a biopharmaceutical, the development process is the same, as illustrated by Figure 1.

Figure 1. FDA's drug development overview, applicable to all drug products, including biopharmaceuticals; obtained from FDA website: www.fda.gov/cder/handbook/index.htm

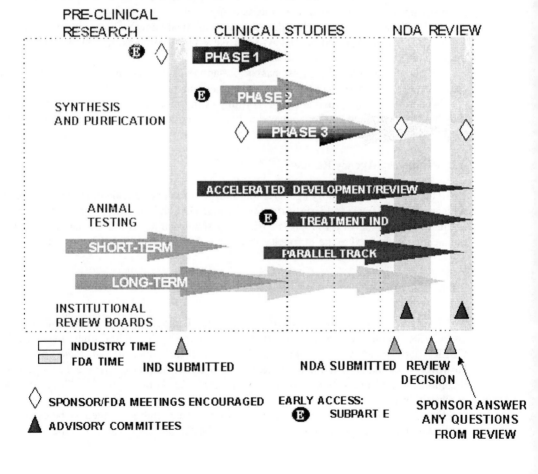

The sponsor must first demonstrate that their product is reasonably safe for use in initial, small-scale human clinical studies. This is typically done during the preclinical stage using laboratory animal testing.

An investigation new drug (IND) application is submitted to the FDA to request initiation of human clinical studies.

Phase 1 clinical studies include the initial introduction of the investigational product into humans and are primarily designed to evaluate the safety of the product and adverse effects with increasing doses.

Phase 2 clinical studies include the early controlled studies conducted in humans to obtain some preliminary data on the effectiveness of the investigational product in a particular indication using a specified dosing scheme.

Phase 3 clinical studies are expanded, well-controlled studies in humans to gather the necessary statistical information about effectiveness and safety of the investigational product that is needed to evaluate the overall benefit-risk relationship.

This drug development process for a new product can take up to 16 years according to recent estimates by the US Food and Drug Administration (FDA).[5] It is for this reason that a number of companies carry out hybrid clinical trials (e.g., Phase 1/2 and Phase 2/3) to try to reduce the total time and resources required. The FDA has been open to this and they have devised various mechanisms that can speed the drug development process, from streamlining their review to accelerated approval.

If the Phase 3 clinical studies demonstrate statistical benefit, the company submits a Biologics License Application (BLA) or a New Drug Application (NDA), depending upon which center of the FDA is involved.

The company, typically, must pass a pre-approval inspection (PAI) and have their clinical study outcome discussed in an advisory committee.

When the FDA is satisfied that everything is in order, they notify the company that they have been granted market approval to being the commercial launch of their product.

3.2. Regulatory Agency Review

The two major regulatory agencies that most biopharmaceutical companies must deal with are the U.S. Food and Drug Administration (FDA) and the European Agency for the Evaluation of Medicinal Products (EMEA). It should be noted that the name EMEA is planned to be changed to the European Medicines Agency (EMA) in the near future.

3.2.1. U.S. FDA

Two centers of the FDA – CBER (Center for Biologics Evaluation and Research) and CDER (Center for Drug Evaluation and Research) – regulate biopharmaceuticals. CBER, until recently, has traditionally regulated the majority of these products. However, it should be noted that CDER has always regulated biopharmaceuticals that are recombinant hormones (such as human insulin and human growth hormone) and combination products that consist of both a biopharmaceutical and a drug component (such as gemtuzumab ozogamicin which is a monoclonal antibody linked to the cytotoxic antitumor antibiotic calicheamicin). In fact, CDER was responsible for the first two biopharmaceuticals approved by the FDA: human insulin in 1982 and human growth hormone in 1985.

From a CMC perspective, does it matter which FDA center regulates your biopharmaceutical? A decade ago the answer would have been an emphatic 'yes', because the differences were huge (e.g., the need for both a product license and an establishment license from CBER; the absence of an annual report mechanism and lack of any tiered approval approach for process changes in CBER; the use of CBER Headquarters staff for inspections; etc.). However over the past decade, CBER and CDER have made significant progress in 'harmonizing' their regulation effort of biopharmaceuticals, even to the point where many were questioning why there were two centers at FDA involved.

Therefore, it was not much of a surprise when the FDA announced in late 2002 that the regulatory review and compliance control of most biopharmaceuticals were being moved from CBER to CDER. The categories of products transferred to CDER in 2003 were:

- Monoclonal antibodies intended for in vivo therapeutic use

- Cytokines, growth factors, enzymes, immunomodulators, and thrombolytics

- Proteins intended for therapeutic use that are extracted from animals or microorganisms, including recombinant versions of these products

- Other non-vaccine therapeutic immunotherapies

Today, CBER regulates only a few biopharmaceutical products:

- Gene therapy involving vectors

- Recombinant vaccines

- Blood clotting factors

3.2.2. EMEA

In Europe, each individual country is currently responsible for approving clinical trials within their control. The new Clinical Trial Directive (2001/20/EC) should make this less cumbersome to start clinical trials in Europe, but implementation of this Directive will not occur until at least mid-2004. Once ratified by the member countries, this Directive will provide a centralized procedure for clinical trial applications.

The European Agency for the Evaluation of Medicinal Products (EMEA), formally operational in 1995, regulates market approval of biopharmaceuticals. All biopharmaceuticals are required to go through the centralized procedure, with review and recommendation to approve through the Committee for Proprietary Medicinal Products (CPMP). A biopharmaceutical company submits a Marketing Authorization Application (MAA). According to EMEA guidelines, the entire review process should take no longer than 300 days (Figure . However, if questions are raised during the evaluation that require additional information from the biopharmaceutical company, the "clock" stops until questions are resolved to the rapporteur's satisfaction.

Figure 2. EMEA's centralized approval process for a biopharmaceutical; obtained from the European Commission website: http://pharmacos.eudra.org/F2/eudralex/vol-2/home.htm

DAY	ACTION
1	Start of the procedure
70	Receipt of the Assessment Reports from Rapporteur and Co-Rapporteur by CPMP members and EMEA. EMEA sends Rapporteur and Co-Rapporteur Assessment Report to the applicant making it clear that it only sets out their preliminary conclusions and that it is sent for information only and does not yet represent the position of the CPMP.
100	Rapporteur, Co-Rapporteur, other CPMP members and EMEA receive comments from Members of the CPMP.
115	Receipt of draft list of questions (including the CPMP recommendation and scientific discussion) from Rapporteur and Co-Rapporteur by CPMP members and EMEA.
120	CPMP adopts the list of questions as well as the overall conclusions and review of the scientific data to be sent to the applicant by the EMEA. **Clock stop.** At the latest by Day 120, adoption by CPMP of request for GMP / GCP inspection, if necessary (Inspection procedure starts).
121*	Submission of the responses, including revised SPC, labelling and package leaflet texts in 13 languages, and restart of the clock. Submission of mock-ups in colour for each strength/form in the smallest pack-size covering all EU official languages, Norwegian and Icelandic and language combinations.

DAY	ACTION
150	Joint response Assessment Report from Rapporteur and Co-Rapporteur received by CPMP members and the EMEA.EMEA sends joint Assessment Report to the applicant making it clear that it only sets out their preliminary conclusions and that it is sent for information only and does not yet represent the position of the CPMP. Where applicable, Inspection to be carried out
170	Deadline for comments from CPMP Members to be sent to Rapporteur and Co-Rapporteur, EMEA and other CPMP Members.
180	CPMP discussion and decision on the need for an oral explanation by the applicant. If oral explanation is needed, the clock is stopped to allow the applicant to prepare the oral explanation. Submission of final inspection report to EMEA, Rapporteur and Co-Rapporteur by the inspections team (at the latest by Day 180).
181	Restart the clock and oral explanation (if needed).
181 to 210	Final draft of English SPC, labelling and package leaflet sent by applicant to the Rapporteur and Co-Rapporteur, EMEA and other CPMP members
By 210	Adoption of CPMP Opinion + CPMP Assessment Report (and timetable for the provision of revised translations)

DAY	ACTION
215 at the latest	Applicant provides the CPMP members with SPC, package leaflet and labelling in the 13 languages. A copy of the cover letter is to be sent to the EMEA for information.
225	Preparation by the applicant of final revised translations of SPC, labelling and package leaflets taking account comments received from EMEA and CPMP.
230 at the latest	Applicant provides EMEA with final translations of SPC, package leaflets and labelling in the 13 languages. Revised full colour mock-ups covering all countries should also be submitted.
By 240	CPMP Assessment Report to be transmitted to the applicant. Transmission of Opinion in all EU languages to applicant, Commission, Member States and Norway and Iceland.
By 300	Finalisation of EPAR in consultation with Rapporteur, Co-Rapporteur, CPMP and applicant (the latter for confidentiality aspects).

4. CMC REGULATORY COMPLIANCE TRACK RECORD

Manufacturers of drugs and biologics, as well as biopharmaceuticals, have not always met CMC regulatory compliance, with the ensuing result of potentially harming patients.

4.1. Drugs and Biologics

Drugs and biologics, while offering tremendous therapeutic benefit, have had a checkered CMC regulatory compliance past, and frequently major disasters have led to increased federal regulations:

- The tragic deaths in 1901 of 13 children in St. Louis who died of tetanus after receiving diphtheria antitoxin from a tetanus-infected horse moved Congress to enact the 1902 Biologics Control Act, which first authorized federal regulation of biologics.

- The death of 107 people, many of them children, in 1937 who had taken elixir of sulfanilamide erroneously mixed with a poisonous solvent moved Congress to enact the 1938 Food, Drug and Cosmetics (FD&C) Act, which expanded FDA's power to regulate drugs and biologics.

- In the early 1950's, a major polio epidemic swept the United States, but Jonas Salk developed an inactivated polio vaccine that was demonstrated to be safe and effective. However, through a tragic set of circumstances known as the "Cutter Incident", more than 100 people contracted polio, and 10 died, from the use of two batches of the polio vaccine. During their production, wild polio virus was not in sufficient contact with formalin to inactivate all virus present. The Cutter lots had no more than 1 infectious

polio virus unit/dose.   On May 7, 1955, the US Surgeon General
recommended that all polio vaccinations be suspended until a thorough
inspection of each manufacturing facility and review of the procedures for
testing vaccine safety had been completed.[6]

- Another surprise with the Salk polio vaccine appeared in 1960.   The
  monkey kidney cells used to make the polio vaccine were found to be
  contaminated with a monkey virus, SV40, a virus that could cause rare
  tumors in hamsters.   More than 98 million people in the US and hundreds
  of millions worldwide had been exposed to the potentially contaminated
  vaccine before vaccine manufacturers were able to eliminate SV40 from
  their cells.   The National Cancer Institute completed a study in 1963
  concluding that children who had received infected vaccine faced no
  increased cancer risk; and in 1981 they concluded that exposed children,
  now teenagers, faced no increased cancer risk.   However, in 2002, SV40
  was spotted in non-Hodgkin's lymphoma patients and other cancer types
  reawakening the debate.[7]

- FDA became aware in the early 1960's of reports of young males,
  reportedly being administered a vitamin preparation, developing female
  characteristics.   FDA inspections revealed manufacturing practices
  suspected of causing contamination with the steroid, diethylstilbestrol.   In
  addition,   these   manufacturing   practices   provided   evidence   of
  contamination of certain drug preparations with penicillin.   Since it was
  difficult at that time to determine residual diethylstilbestrol, various
  samples were collected for the analysis of penicillin content to support the
  evidence of poor manufacturing practices; the laboratory findings from the
  Division of Antibiotics confirmed the presence of penicillin in several non-
  antibiotic preparations.   Amendments to drug regulations for current
  GMPs were published in 1965 for the control of cross-contamination by
  penicillin.[8]

4.2. Biopharmaceuticals

Biopharmaceuticals are still new, but that doesn't mean that they are without
concern.

Excitement surrounded many gene therapy clinical trials until 1999.   In September
1999, a patient died from a reaction to a gene therapy treatment at the University of
Pennsylvania's Institute of Human Gene Therapy.   Jesse Gelsinger, an 18-year old,
suffered from ornithine transcarboxylase deficiency (OTCD), a broken gene that prevents
the liver from making an enzyme needed to break down ammonia. He entered a clinical
trial to receive the OTCD gene package in a replication-defective adenovirus. To reach
the target cells in the liver, the adenovirus was injected directly into the hepatic artery
that leads to that organ. Four days later he was dead. No one is sure why the gene therapy
treatment caused his death, but it appears that his immune system launched a raging

attack on the adenovirus carrier. In the aftermath of his death, the FDA launched several investigations of the University of Pennsylvania studies and others. The inquiries provided disappointing news: gene therapy researchers were not following all of the federal rules requiring them to report unexpected adverse events associated with the gene therapy trials; in addition, they were providing inadequate oversight of manufacturing and record keeping; worse, some scientists were asking that problems not be made public. And then came the allegations that there were other unreported deaths attributed to genetic treatments, at least six in all. The fallout from this tragedy not only included loss of public confidence but also increased CMC demands by FDA from companies desiring to pursue gene therapy clinical trials.[9]

Then, a major concern struck the retrovirus gene therapy programs in 2003 when two children developed leukemia, potentially as a result of an otherwise successful treatment with gene therapy for a condition which is frequently otherwise fatal: X-linked severe combined immunodeficiency syndrome (X-SCID, also known as "bubble baby syndrome"). A temporary hold was placed on all retroviral gene therapies and after advisory committee review, recommendations were made that clinicians warn their patients that retrovirus therapy can cause cancer.[10]

Manufacturers of both bioengineered animals and plants must adhere to a strict set of containment rules to prevent these recombinant organisms from compromising the animal and human food sources. Companies that do not adhere strictly to regulatory compliance requirements are treading down a dangerous pathway. The survival of this type of technology to produce biopharmaceuticals will depend on how well they manage this and whether they can maintain public confidence in their abilities to do it.

Concerns about plant based biopharmaceuticals compromising the food supply is becoming a public issue with the use of transgenic plants. Two public incidents have raised concerns. First, in September 2000, DNA fragments from StarLink corn (corn genetically modified to express the Cry9C pesticidal protein to make the corn more resistant to certain types of insects) were detected in taco shells, and then in November 2000, the FDA issued a public notice that Aventis CropScience had confirmed that StarLink bioengineered protein had been found food-grade corn seed.[11] Much congressional and public discussion resulted over this. Registration to sell StarLink was withdrawn. And more recently, in November 2002, the FDA issued a public notice that a biotechnology firm, ProdiGene, had agreed to strengthen the controls on any bioengineered plant it grew that would be used in current and future clinical studies. At issue was the discovery of a small number of their volunteer genetically engineered corn stalks harvested in a field of soybeans in Nebraska and subsequently commingled with approximately a half of million bushels of soybeans. The entire lot of soybeans was subsequently destroyed to prevent them from entering the human or animal food supply.[12]

Transgenic animals have the same issue in potentially compromising the food supply. In February 2003, the FDA disclosed that they had determined that bioengineered piglets were not properly disposed of and had possibly entered the food supply.[13] It should be noted that the FDA could not verify the researchers assertion that the offspring of the transgenic pigs did not inherit the inserted genetic material from their parents because the researchers did not conduct sufficient evaluation or keep sufficient records.

However, the good news for biopharmaceuticals to date is summarized by CBER in its 2002 review of its rapid product approval performance:

"There has never been a need to recall an Office of Therapeutics Research and Review-approved biotechnology drug due to safety concerns."[14]

It is the wish of the biopharmaceutical industry that such a safety profile continues. CMC regulatory compliance of biopharmaceuticals is necessary to protect our patients.

Are Biopharmaceuticals Really Different?

Few drug discovery stories have offered researchers as many chances for dismissive belief as the one William B. Coley launched with his bacterial lysate treatments of cancer. When Coley learned of a cancer survivor who coincidentally had suffered sever skin infection caused by *Streptococcus pyogenes*, he wondered if the bacteria had caused the patient's tumour to regress. In the 1890's he began injecting cancer patients with crude *Serratia marcescens* bacterial preparations that became known as Coley's toxins, with claims that some achieved lengthy remission. Many others who tried to repeat the results failed. Coley endured decades of derision from the American Chemical Society.

Coley's toxin – a possible precursor to today's anti-cancer biopharmaceuticals now known to induce fever (recombinant IFN, IL, TNF, etc.)

1. PERCEPTION OR REALITY?

It can be quite a shock for the person who has been working in the Chemistry, Manufacturing and Controls (CMC) area with traditional pharmaceutical drugs to move over and start working with biopharmaceuticals.

1.1. Five Questions Frequently Asked

Three of the first questions that are often raised by new people about biopharmaceutical operations are:

1. Why does the cost have to be so high for operating and supporting the manufacturing facilities, especially the air handling and water utility systems'

2. 'Why is it necessary to have such a large number of staff, especially Quality staff; to support the manufacturing operations?'

3. 'Why is so costly to validate a biopharmaceutical process, especially the purification steps?'

Then when their attention is turned toward biopharmaceutical product inventory, two additional questions that are often raised are:

4. 'Why does it take so long, possibly even months, for the product to be released by Quality Assurance?'

5. 'Why can't we increase our product shelf-life by extrapolating from the existing stability data?'

1.2. Bottom Line Question

These five questions can be boiled down to the single critical question: Are the differences that are observed between biopharmaceuticals and chemically-synthesized drug products real or just perceived? From those of us who have worked in this industry for so many years, it is surprising that such a question would even be raised today. But then again, there are still people in the industry that feel that the differences are being made up either by the biopharmaceutical CMC consultants (for their job security) or by their own Quality Assurance groups (for their power trip). Unfortunately, this is not a healthy attitude for a company in this business to take.

But if there are real differences, and there are, it is most important that a company recognize such differences and properly address from a corporate strategic perspective the additional CMC regulatory compliance issues for their biopharmaceutical product.

2. REGULATORY AGENCIES SPEAK

The best answer to the question of whether the CMC differences that are observed between biopharmaceuticals and chemically-synthesized drug products are real or perceived comes from the regulatory agencies themselves.

2.1. U.S. FDA

The FDA believes very clearly that there is a difference between biopharmaceuticals and chemically-synthesized drugs:

- FDA answering why more CMC information on container closure systems is required for biopharmaceutical products than other drugs:

  "... there is greater potential for adverse effects on the identity, strength, quality, purity, or potency of biologics and protein drug products during storage or shipping."[15]

- FDA discussing the amount of CMC description of the method of preparation of the drug substance required for their review of Phase 1 INDs:

  "More information may be needed to assess the safety of biotechnology-derived drugs ...."[16]

- FDA discussing the agenda for CMC meetings with the agency:

  "The discussion of safety areas for conventional synthetic drugs is typically brief. For certain type of drugs, such as biotechnology drugs ... it may be appropriate to discuss the CMC information in more detail."[17]

- FDA explaining the scope of their comparability protocol guidance:

  "This guidance applies to comparability protocols that would be submitted in NDAs and ANDAs ..., except for protein products. A separate guidance will address comparability protocols for proteins as well as for peptide products outside the scope of this guidance."[18]

- FDA announcing the plan for moving some biopharmaceuticals from CBER to CDER:

  "It was again reiterated that under the new structure the biologic products transferred to CDER will continue to be regulated as licensed biologics."[19]

2.2. EMEA

EMEA believes that there is a difference between biopharmaceuticals and chemically-synthesized drugs:

- EMEA explaining its note for guidance on process validation:

    "The note for guidance ... It is not intended to apply to products of biotechnological ... since these processes are themselves very complex in nature and have an inherent variability which generally require the submission of more extensive validation data."[20]

- EMEA introducing its concept paper of PK and PD principles for proteins

    "An increasing number of new Marketing Authorization Applications for medicinal products concern peptides or proteins indicated for therapeutic or diagnostic use. The pharmacokinetic behavior of such compounds is generally different from that of "conventional" molecules and, consequently, alternative issues have to be considered when the development process is designed and study results are interpreted."[21]

2.3. ICH

The International Conference on Harmonization (ICH; tripartite agreements reached between the FDA, EMEA and the Japanese Ministry of Health, Welfare and Labor) believes there is a difference:

- ICH Q7A – GMPS for Active Pharmaceutical Ingredients, explaining why there is a separate section for biopharmaceuticals:

    "GMP principles for the manufacture of APIs for use in human drug products are not adequate for APIs manufactured by cell culture or fermentation using recombinant organisms."[22]

- ICH Q6A – Specifications for Chemically Synthesized Drugs, explaining the scope of this document:

    "This guideline applicable to drugs of synthetic chemical origin is not sufficient to adequately describe specifications of biotechnological and biological products."[23]

In fact, because of the difference between biopharmaceuticals and chemically-synthesized drugs, ICH has several documents that cover the same CMC topics but are applicable either only to biopharmaceuticals or only to chemically synthesized drugs:

CMC Topic	Biopharmaceutical Drugs	Chemically-Synthesized Drugs
Stability	ICH Q5C	ICH Q1A
Specifications	ICH Q6B	ICH Q6A

3. THREE UNIQUE CMC CHALLENGES FOR BIOPHARMACEUTICALS

With all of these comments from the regulatory agencies, there has to be a real difference between biopharmaceuticals and chemically-synthesized drugs! As a result, biopharmaceuticals should have unique CMC challenges. In fact they do, and the CMC challenges come from three areas:

1. The use of living recombinant organisms

2. The products themselves

3. The impact of the manufacturing process.

3.1. The Use of Living Recombinant Organisms

Biopharmaceuticals are the result of applying genetic engineering to living cells, whether it be bacteria cells, yeast cells, mammalian cells, viruses, whole animals or whole plants. These living hosts can provide an opportunity for amplification of various types of adventitious agents (e.g., bacteria, fungi, mycoplasma and viruses) during their production. The amplified adventitious agent can overwhelm the standard control procedures of the manufacturer and pose a safety risk to the patients intended to be helped by the biopharmaceutical. Biopharmaceuticals cannot be terminally sterilized which could have provided a barrier downstream of cell culturing.

Adventitious agent contamination is minimized by stringent adherence to current good manufacturing practices (CGMPs) and proper containment procedures, as well as careful, thorough quality controls and testing. But adventitious agent contamination cannot be entirely eliminated from a manufacturing process so contamination does occasionally happen. Should amplification of the agent occur, it is most important for the biopharmaceutical manufacturer to be vigilant for such an event and prepared to do what is necessary to protect the facility, the equipment and downstream processes from the adventitious agent. Ultimately, the manufacturer must demonstrate that the product is free of adventitious agents and thus poses no risk to the patient.

Risk management for biopharmaceuticals is an ongoing process because of working with these living organisms. And it requires staying current with new risks. Who was concerned in the early 1980's with the risk of bovine spongiform encephalopathy (BSE) being possibly present in the bovine components that we were using in the cell culture process? Who was concerned before the year 2002 about the risk of West Nile Virus being present in the human blood-derived components that could be used in their manufacturing processes?[24] Who was concerned before the year 2003 about the risk of the causative virus for Severe Acute Respiratory Syndrome (SARS) being present in these same human blood-derived components?[25]

3.2. The Products Themselves

Biopharmaceuticals, when viewed from a two-dimensional perspective, are really simple molecules composed of a handful of amino acids linked together in various lengths and combinations to form the primary protein chains, and many times having a handful of carbohydrates linked together in various lengths and combinations to specific amino acids on the chain (i.e., glycosylation). However, various manufacturing systems starting from the same gene can yield the same biopharmaceutical but with differing post-translational modifications (e.g., glycosylation patterns and protein heterogeneity). These changes may not always be readily detected by available analysis tools.

But, when biopharmaceuticals are viewed from a three-dimensional perspective, the major CMC challenges can be understood. For biopharmaceuticals, the maintenance of its three dimensional form, its molecular conformation, is critical, as it can have dramatic impacts on either the biological activity of the molecule, or its immunogenicity, or both, when administered to the patient. But the maintenance of molecular conformation is dependent on both covalent forces and non-covalent forces, which are often hard to control during the manufacturing process and also when the product sits in a container on the shelf.

Furthermore, the ability to distinguish one protein product from another is not trivial, after all, for the most part proteins have the same amino acids and carbohydrate components present. This places extra pressure upon the manufacturer to ensure that products are not mixed during manufacture or subsequent handling. Identity for a biopharmaceutical is more than showing that protein is present; it involves demonstrating that the specific protein is present. A simple 1 hour long infrared spectrophotometer fingerprint identity test for a chemically-synthesized drug can become a 1 week long peptide mapping fingerprint identity test for a biopharmaceutical.

3.3. The Impact of the Manufacturing Process

New technologies are often met with ultra-conservative CMC regulatory compliance controls. In the past when biopharmaceuticals were new and their technology and manufacturing processes were new, the FDA was treading cautiously and took the position that 'the process was the product'. CBER required two licenses for each biopharmaceutical feeling the need to control both the product (through a product license application, PLA) and the manufacturing process in its facility (through an establishment license application, ELA). All biopharmaceutical process changes required FDA pre-approval prior to implementation.

Today, because of considerable experience with the biopharmaceutical products and their manufacturing processes, the FDA has now taken the position that 'the process impacts the process'. The FDA now requires a single license (a biologic license application, BLA, or a New Drug Application, NDA) and even now permits some biopharmaceutical process changes to occur without prior approval or within a 30-day window for pre-approval by the agency.

But this does not mean that there is not considerable concern about the control of the manufacturing process. The process can impact the molecular conformation as stated above, but the manufacturing process can also directly alter the product itself. During a bacterial fermentation process with recombinant *E. coli*, if the culture medium becomes methionine-depleted, the biopharmaceutical produced can have an incorrect amino acid, norleucine, incorporated into the molecule where methionine should be.[26] During a mammalian cell culturing process, if the process is not properly controlled, the resulting biopharmaceutical can have different carbohydrate isoforms present that can impact the *in vivo* activity and/or serum half-life of the molecule.[26] Additional concerns about the process impacting the product are discussed in Chapter 10.

There is considerable discussion today about whether the FDA will move their opinion about biopharmaceuticals to that currently accepted for chemically-synthesized drugs; that is, 'the process can be independent of the product'. While there are generic chemically-synthesized drugs (using an abbreviated NDA, and not being required to perform clinical trials prior to FDA approval), there is no equivalent for biopharmaceuticals. From this author's experiences and based on further CMC discussion in this book, it is difficult to see such a move for biopharmaceuticals. The risk that would be incurred in moving to 'generic biologics' can best be summed up by the following statement:

> "Safe and effective biological products can be assured only by extensive characterization of each drug substance and product, a well-defined and validated manufacturing process, and appropriate clinical trials to establish the safety and efficacy for that molecule".[26]

4. CMC MEETINGS WITH THE FDA TAKE ON GREATER IMPORTANCE

Because of the unique CMC challenges for biopharmaceuticals, maintaining an ongoing open dialogue with the regulatory agencies is critical for moving the biopharmaceutical through the clinical development stages and into the marketplace. The FDA in announcing its new strategic plan for beyond 2002 identified poor communication as a significant factor in causing unnecessary delays in new product approvals:

> "In addition, the findings of these retrospective analyses of causes for delays in approval emphasize the importance of communication between FDA and the sponsor during the product development phase so that various deficiencies can be addressed prior to submission of the application or avoided altogether by

better drug development practices. Clear understandings of agency expectations and timely communications between FDA and application sponsors can increase the likelihood that a submitted application contains the necessary information for timely approval in the first round. For example, insufficient FDA-industry interactions may result in late identification of important problems and other avoidable concerns that lengthen the time and cost of product development when discovered late in the development or review process and necessitate further data development and delays in product marketing approval."[27]

The FDA has identified 3 CMC specific meetings that could be of value to a company during the evolving stages of clinical trial development for a biopharmaceutical (Figure 3).

Figure 3. CMC specific meetings with FDA, of special importance to biopharmaceuticals

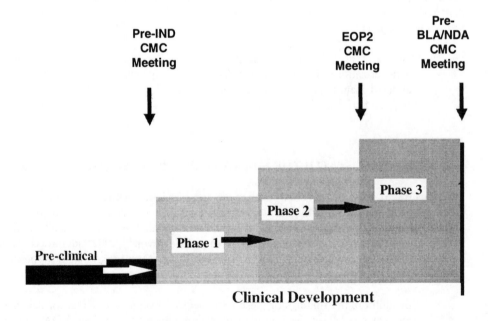

4.1. CMC Communication with FDA is Critical

It may not be clear to all why CMC specific meetings with the regulatory agencies are of extra importance to biopharmaceutical companies. There are still some companies that believe that they know more than the regulatory agency reviewers and they view their input as possible interfering with the direction that the company wants to move down: 'Why discuss that with the FDA, they may want us to do something differently?' and 'Isn't it a better use of our limited resources to focus our resources on the clinical meetings instead?' When these questions arise, you may need to convince your senior management of the value of these meetings.

A good start is to bring to their attention the comments from the FDA about the need for these CMC specific meetings, especially for biopharmaceuticals:

- Prior to Filing the IND: "The discussion of safety areas for conventional synthetic drugs is typically brief. For certain type of drugs, such as biotechnology drugs ... it may be appropriate to discuss the CMC information in more detail."[17]

- End of Phase 2: "The CMC portion of the EOP2 meeting is a critical interaction between the sponsor and the chemistry review team to ensure that meaningful data will be generated during phase 3 studies. The goal is to identify potential impediments to further progress at an early stage, thus reducing the number of review cycles for the proposed marketing application. Although the EOP2 meeting is important for all drugs, it is particularly important for new molecular entities, biotechnology drugs...."[17]

Also, it can be helpful to remind senior management that the FDA has continually encouraged companies to discuss CMC issues with them before they finalize their plans:

- "FDA urges manufacturers to consult with FDA prior to implementing changes that may result in comparability testing, in order to avoid delay in review of applications."[28]

- "Sponsors considering novel expression systems not specifically covered by guidance documents are encouraged to consult with CBER."[29]

It should be obvious that no meeting should be held unless there is a genuine reason. For biopharmaceuticals there are at least three primary purposes for these CMC specific meetings:

1. To address outstanding questions and scientific issues that arise during the course of the biopharmaceutical's development

This is especially important for new types of biopharmaceuticals or new manufacturing technologies that produce the biopharmaceutical product, where there may be limited existing CMC regulatory compliance guidance published.

2. To aid in the resolution of problems

Every biopharmaceutical will encounter problems or challenges, and with the regulatory agencies' willingness to provide direction, they can provide assurance that the approach to be taken to resolve the problem will be satisfactory. Also, they have been known to suggest additional workable solutions that a company might consider for resolving the problem.

3. To facilitate evaluation by the regulatory agency of the biopharmaceutical

The company, the regulatory agency and the patients all win when a new biopharmaceutical product is approved for the market. To avoid delays in approval, especially if expedited or accelerated drug approval is granted, a company's CMC issues must be handled effectively. Early interaction with the regulatory agencies is essential, as they ultimately have to accept the final CMC strategy and documentation.

4.2. Preparing for the CMC Meeting

But keep in mind, when preparing for a CMC specific meeting, follow the guidance provided by the FDA (Table 4). The FDA has provided detailed instructions on how to request a meeting, how to prepare the CMC briefing document and how to focus the meeting on specific CMC questions. Not following these administrative details can delay the scheduling of a meeting which could impact the forward progress of your product. A background briefing document is mandatory prior to the meeting. The goal of this submission should be to provide an easy to read and easy to understand account of the issues to be discussed.

Table 4. Instructions from the FDA on what companies should include in a request letter when asking for a CMC specific meeting [30]

The product name and IND number, if already assigned

Chemical name and structure

The type of CMC meeting being requested (Type A for stalled drug development programs; Type B for a standard meeting to review drug development progress; and Type C for meetings such as facility design, general product issues or general development)

A brief statement of the purpose for the meeting
- This should include a discussion of the types of studies or data that the sponsor or applicant intends to discuss at the meeting
- For new products, this should include a description and developmental status of the product, as well as its proposed indication

A listing of the specific objectives or outcomes that the requester expects

A preliminary proposed agenda, including estimated times needed for each agenda item and designated speaker(s)

A draft list of specific questions, as comprehensive and precise as possible

A list of all individuals (including titles) who will attend the proposed meeting from the sponsor's or applicant's organization and consultants

A list of requested participants or disciplines to be represented from the Center

The approximate time that a background package for the meeting will be sent to the Center (i.e., x weeks prior to the meeting), consistent with the type of meeting planned (Type A, at least 2 weeks; Type B, at least 1 month; and Type C, at least 1 month in advance)

Suggested dates and times (i.e., morning or afternoon) for the meeting

4.3. Pre-IND Meeting

The first key CMC meeting between the sponsor and the FDA is the pre-investigational new drug (pre-IND) meeting, which takes place prior to filing the IND to initiate Phase 1 clinical trials. For many first-product biopharmaceutical companies, this might be their first exposure to the regulatory agency.

CMC issues need to be discussed if they could lead to a potential clinical hold, and this especially applies to biopharmaceuticals. To have a rewarding dialogue and obtain the necessary agency guidance, the FDA requires a written briefing document to be submitted to them prior to the meeting (usually 4 weeks in advance). Table 5 presents some of the CMC issues that could be focused on during the pre-IND meeting for a biopharmaceutical.   Not that all of these areas need to be discussed but if there are issues with your biopharmaceutical manufacturing process or the biopharmaceutical itself, this is a great opportunity to seek FDA guidance.

Table 5. CMC issues recommended by the FDA for discussion at a pre-IND CMC meeting[17]

Physicochemical and biological characterization of the product

Manufacturers

Source and method of manufacturing

Removal of toxic reagents

Adequacy of cell bank characterization

Potential contamination of cell lines

Removal or inactivation of adventitious agents

Potential antigenicity of the product

Formulation, especially if novel excipients are used

Quality controls (e.g., identity, assay, purity, impurity profile)

Sterility (e.g., sterilization process, release sterility and endotoxin testing)

Stability information

Linkage of pharmacology and/or toxicity lots to clinical trial lots

From my experience, the regulatory agency's interest in having a pre-IND meeting increases if the biopharmaceutical is manufactured using either a new cell line (especially if there is an endogenous virus present) or a newer technology (e.g., transgenics). Interest also increases if the biopharmaceutical product has challenges (e.g., purification issues or unanticipated structural issues with the biopharmaceutical). Always keep in mind that the 'newness' of either the biopharmaceutical product or its manufacturing process will trigger interest from the FDA, and the unknowns will most likely then trigger a conservative regulatory agency response. Assurances will have to be provided that the company will be able to exercise adequate CMC control during the manufacturing and release of the biopharmaceutical for clinical trials. Following up on these assurance will bring confidence to the FDA about the company's commitment to quality and compliance.

Less one think that the CMC at Phase 1 is not considered significant by the regulatory agencies, the FDA has stated that if they believe that there are any reasons that the manufacturing and controls for the clinical trial product presents unreasonable health risks to the patients, that they will delay or suspend all or part of the clinical work requested in the IND. Specifically, they include the following CMC examples as unreasonable health risks to the patients[16]:

- Product made with unknown or impure components

- Product possessing chemical structures of known or highly likely toxicity

- Product that cannot remain chemically stable throughout the testing program proposed

- Product with an impurity profile indicative of a potential health hazard or an impurity profile insufficiently defined to assess a potential health hazard

- Poorly characterized master or working cell bank

4.4. End of Phase 2 (EOP2) Meeting

The second key CMC specific meeting between the sponsor and the FDA is the end of Phase 2 (EOP2) meeting, which usually takes place prior to initiating the pivotal clinical trial. Although the EOP2 meeting is important for all pharmaceuticals, the FDA considers this meeting especially important for biopharmaceuticals to ensure that potential impediments to further progress at an early stage are identified.

From my experience, biopharmaceutical companies are not committing adequate resources to properly prepare the written briefing document that must be submitted ahead of time to the agency. The easier that it is for the FDA reviewers to understand the concerns, the more confidence that a company will have in the final agreements reached with the agency. A list of CMC questions should be included in the briefing document and they should be specific, comprehensive and precise as possible to identify the critical issues. Also, sufficient CMC background should be included to allow the regulatory agency to address the specific questions. The company should understand that this CMC meeting provides them the opportunity to present their results of the biopharmaceutical development program to date for evaluation by the agency. This meeting provides the company the ability to identify and resolve, if possible, any specific safety or scientific issues or problems prior to initiation of the Phase 3 studies. This meeting provides the FDA the opportunity to identify additional CMC information that the company might need to generate important to support a future marketing application. In other words, this is a very critical meeting for both the FDA and the company, and the company should prepare accordingly. The FDA requires the written briefing document to be submitted to them prior to the meeting (usually 4 weeks in advance).

Table 6 presents some of the CMC issues that could be focused on during the EOP2 meeting for a biopharmaceutical. It is important that the company use this opportunity of discussion with the agency to resolve any major CMC issues that could ultimately delay product approval. This is not the time for a company to hold back and be shy about its CMC issues!

Table 6. CMC issues recommended by the FDA for discussion at an EOP2 CMC meeting[17]

Unique physicochemical and/or biological properties of the product

Adequacy of physicochemical and biological characterization studies (e.g., peptide map, amino acid sequence, disulfide linkages, higher order structure, glycosylation sites and structures, other post-translational modifications, and plans for completion, if still incomplete)

Coordination of all manufacturing activities, including full cooperation of Drug Master File (DMF) holders and other contractors and suppliers in support of the planned BLA/NDA

Starting material designation

Adequacy of cell bank characterization (update from Phase 1/Phase 2, plans for completion, if still incomplete)

Removal or inactivation of adventitious agents (update form phase 1, where applicable)

Removal of product- and process-related impurities (e.g., misfolded proteins, aggregates, host cell proteins, nucleic acids)

Bioactivity of product-related substances and product-related impurities relative to desired biopharmaceutical

Approach to specifications (i.e., tests, analytical procedures and acceptance criteria)

Bioassay (e.g., appropriateness of method, specificity, precision)

Approach to sterilization process validation and/or container closure challenge testing, where applicable

Appropriateness of the stability protocols to support phase 3 studies and the planned BLA/NDA

Linkage between formulations and dosage forms used in preclinical, clinical and pharmacokinetic and pharmacodynamic studies, and formulations planned for the BLA/NDA

Major CMC changes, including site changes, anticipated from phase 2 through the proposed BLA/NDA, ramifications of such changes, and appropriateness of planned comparability and/or bridging studies, if applicable

Environmental impact considerations, if pertinent

Identification of any other CMC issues, including manufacturing site, which pose novel policy issues or concerns, or any other questions, issues or problems that should be brought to the attention of the Agency or sponsor

From my experience, the following are the top 5 CMC hot topic issues for a biopharmaceutical product that the FDA wants to discuss at the EOP2 meeting:

- Sufficient product characterization

  Does the company really understand their molecule?

- Product comparability after process changes

  Will the product really be the same after future process changes?

- Management of outsourced CMC

  Who's in charge at the contract manufacturer(s)

- Bioassay (Potency)

  What is the biological function assay?

- Company's approach to justifying specifications/stability

  Does the company have a strategy?

4.3. Pre-BLA/NDA Meeting

The third key CMC specific meeting between the sponsor and the FDA is the pre-biologics license application (BLA) or pre-new drug application (pre-NDA) meeting, which usually takes place about 6 months prior to the submission of the marketing authorization. With respect to CMC requirements, the meeting is to ensure that the proposed dossier submission is well-organized and complete. The goal of the meeting is to resolve any CMC problems that could cause a refusal-to-file recommendation by the agency or hinder their review process. As with the other CMC meetings, the FDA requires a written briefing document to be submitted to them prior to the meeting (usually 4 weeks in advance).

Table 7 presents some of the CMC issues that could be focused on during the pre-BLA/NDA meeting for a biopharmaceutical. Note, that this meeting provides an opportunity for the FDA to obtain assurances from the company that it will have all of the required CMC content in the submitted dossier and that the company is ready for a pre-approval inspection at the time of filing.

Table 7. CMC issues recommended by the FDA for discussion at a pre-BLA/NDA CMC meeting[17]

Discussion of the format of the proposed marketing authorization, including whether an electronic submission will be provided

Confirmation that all outstanding issues discussed at the EOP2 meeting or raised subsequently will be adequately addressed in the proposed filing

Assurance that all activities in support of the filing have been coordinated, including the full and timely cooperation of DMF holders or other contractors and suppliers

Discussion of the relationship between the manufacturing, formulation, and packaging of the drug product used in the phase 3 studies and the final drug product intended for marketing, and assurance that any comparability or bridging studies agreed upon at the EOP2 meeting have been appropriately completed

Assurance that the submission will contain adequate stability data in accordance with stability protocols agreed upon at the EOP2 meeting

Confirmation that all facilities (e.g., manufacturing, testing, packaging) will be ready for inspection by the time of the BLA/NDA submission

Identification of any other issues, potential problems (especially those that could lead to a refuse–to-file recommendation), or regulatory issues that should be brought to the attention of the Agency or sponsor

From my experience, one or more of the following three CMC issues are usually the focal point of the pre-BLA/NDA meetings for a biopharmaceutical:

- Can the company document that it has honored its commitments made previously to the FDA?

- Can product comparability be demonstrated (especially if scale up or process changes are planned after the pivotal clinical studies are completed)?

- Does the company understand the extent of full CMC content needed in the filing of the dossier?

It is important that the company use this opportunity of discussion with the agency to resolve any major CMC issues that could ultimately delay product approval. Since the FDA will see the CMC issues later in the submitted dossier or during its pre-approval inspection, this is not the time for a company to hold back and be shy about its CMC issues!

Having full CMC content in the filed BLA/NDA is not a negotiable issue with the agency.   The FDA can issue a refusal-to-file (RTF) for any biopharmaceutical dossier that is incomplete in the following CMC issues[31]:

- Incomplete description of relevant cell banking systems

- Omission of data demonstrating consistency of manufacture

- Incomplete data demonstrating equivalency to clinical trial product when significant changes in manufacturing processes of facilities have occurred

- Failure to describe changes in the manufacturing process, from material used in clinical trials to commercial production lots

And having a dossier RTF'd becomes an issue for public reporting by the company. This is clearly not something that any senior management wants to have to deal with.

5. WHAT ABOUT CMC MEETINGS WITH EMEA?

Unfortunately, the EMEA is not as accessible to biopharmaceutical manufacturers during the clinical development stages as with the FDA. The EMEA relies heavily on published guidances to guide the industry. However, EMEA does have an official procedure where a company can request scientific guidance.[32] But this guidance has the following limitations:

- Scientific advice will be given by the CPMP on questions concerning specific issues relating to the manufacturing of the product

- Scientific advice is restricted to purely scientific issues

- Scientific advice is not binding to the EMEA with regard to any future marketing authorization of the product concerned

- It is not the role of the CPMP to substitute the industries' responsibility in the development of their products

In the United States, we can be thankful for the open door communication policy of our FDA.

6. BIOPHARMACEUTICALS NEED TO BE TREATED DIFFERENTLY

Biopharmaceuticals are truly unique. The use of genetically modified living organisms for production, the uniqueness of the biopharmaceuticals themselves, and the impact of the manufacturing process on the product, all create the CMC regulatory compliance challenges that will be discussed in great detail in the following chapters.

The need for active involvement with the regulatory agencies during the clinical development of these products is paramount. Use every opportunity to obtain FDA guidance. Biopharmaceuticals move faster through the regulatory review process when companies treat FDA reviewers as team members.

The FDA has the major task to keep pace with the development of biopharmaceuticals, and, in my opinion, they have done a remarkable job, especially considering the diversity of biopharmaceuticals that they must regulate. FDA's regulation of these products, as stated by the regulators themselves, "should be based on sound science and good sense".[33] It is the hope of the biopharmaceutical industry that this practice continues.

3

Developing the Corporate CMC Regulatory Compliance Strategy

"There's a way to do it better – Find it."

Thomas A. Edison, directive to a research assistant, c. 1919

1. THREE KEY ELEMENTS FOR A COMPLETE CMC STRATEGY

'We have to move fast because the competition is on our tail; we have limited funds and staff to move our biopharmaceutical through clinical development so we cannot do everything and hire everybody we want; so what CMC must we absolutely have to do now to be successful, and what can we do at a later stage of clinical development?'

Because the CMC challenges seem so overwhelming for a biopharmaceutical, this question, or some version of it, is often asked by senior management in biopharmaceutical companies.

And it is a fair question to ask, since there is so much CMC work that could be done for the products. On the one hand, many biopharmaceuticals fail before or during Phase 3 clinical trials and it would waste considerable corporate resource if too much CMC work was done too soon at earlier clinical development stages. Patients are not benefited if the company runs out of funding and cannot bring the drug products that provide clinical benefit into the market. On the other hand, if not enough CMC work is carried out at earlier clinical development stages, the company may be hard pressed to meet the expectations of the regulatory agencies if the biopharmaceutical quickly demonstrated significant clinical benefit and was granted an expedited review. Patients are not benefited if there are delays in product approval. Therefore, it is important that a biopharmaceutical company be well aware of the risk that they are incurring by the choice of the corporate CMC regulatory compliance strategy that they pursue.

A better question to ask is: 'What CMC must I do to at each stage of clinical development to protect the product, to protect the patient and to protect the corporate reputation with the regulatory agencies?' To address this question, three (3) key elements involved in developing a successful corporate CMC regulatory compliance strategy need to be assessed.

1.1. Element 1: The Broad CMC Scope Must Be Considered

The first key element is to make sure that the strategy considers all aspects of the CMC picture involved with the manufacture, testing and release of the biopharmaceutical product. The full CMC scope includes all of the following:

- Manufacturing Facility: design, operation and maintenance

- Support Utilities: design, qualification, operation and maintenance

- Process Equipment: design, qualification, operation and maintenance

- Process Materials: acceptance criteria and vendor qualifications

- Personnel: levels and competency

- Process: consistency and validation

- Product: characterization and proof of its structure

- Quality Control: release testing, specifications and stability

- Quality Assurance: release and rejection

- Regulatory: compliance with all regulations and guidances, and with all past regulatory agency commitments

At each stage of clinical development, an assessment must be made of what is required at that stage for compliance with good manufacturing practices and from proper design and operation of the facility, the utilities, and the process equipment. The assessment must determine if the manufacturing process is adequately controlled and if there is adequate, trained staff available both in manufacturing and quality. The assessment must also determine if there is an adequate understanding of the biopharmaceutical product available, in terms of its characterization, release specifications and stability profile. And it must determine if all specific CMC commitments made to a regulatory agency are being followed.

A process is only as strong as its weakest link, and that is why the company must consider the broad CMC scope in its assessment. Having the newest manufacturing facility gains little if the manufacturing process is out of control or the operations are not carried out under good manufacturing practices. Having tight release specifications gains little if there is low confidence in the analytical methods used to generate the data. Claiming to follow all FDA guidances helps, but not if a company ignores its other CMC commitments that it made to the agency during meetings or written correspondence.

1.2. Element 2: Any Unique CMC Issues Must Be Addressed

The second key element is to ensure that the corporate CMC strategy addresses any unique challenges for the type of biopharmaceutical in clinical study. For the most part, biopharmaceuticals, regardless of the technology used to produce them, pose very similar CMC issues. Yet there are some CMC differences that must be addressed. Some examples include:

- Cell Based Production Systems

    o Documenting the absence of transmissible spongiform encephalopathies (TSEs) from the wide variety of animal-derivied and human-derived raw materials that could potentially be used in a cell-base production process

    o 'Wedded to stainless steel'

- Gene Therapy

    o Absence of replication-competent virus (RCV) when viral vectors are used to transfer genes into patients

    o Patient-specific products and the need to inject the product into the patient prior to final sterility test results

- Animal Transgenics

    o Avoidance of the use of live attenuated viruses when attempting to vaccinate animals with veterinary vaccines

    o Stringent containment issues to protect contamination of food sources

- Plant Transgenics

    o Natural plant viruses

    o Potential adventitious virus exposure from growing crops in open field (birds, insects, mammals and humans)

> o   Stringent containment issues to protect contamination of food sources

Thus, containment and control will be defined differently for different types of biopharmaceutical products. The corporate CMC regulatory compliance strategy needs to first recognize and then address these differences.

1.3. Element 3: Must Meet Minimum CMC Regulatory Requirements

The third key element of the corporate CMC strategy is to meet the minimum CMC regulatory requirements at each stage of clinical development. The minimum CMC requirements to initiate human clinical studies when filing an Investigational New Drug (IND) amendment at Phase 1 will be much less than the minimum CMC requirements when filing the Biologics License Application (BLA)/New Drug Application (NDA) at the completion of pivotal clinical trials. This graded nature of CMC requirements is clearly indicated by FDA regulations in the Code of Federal Regulations [21 CFR 21 Part 312.23(a)(7)(i and ii)]:

> "Although in each phase of the investigation sufficient information should be submitted to assure the proper identification, quality, purity, and strength of the investigational drug, the amount of information needed to make that assurance will vary with the phase of the investigation, the proposed duration of the investigation, the dosage form, and the amount of information otherwise available.
>
> FDA recognizes that modifications to the method of preparation of the new drug substance and dosage form itself are likely as the investigation progresses.
>
> Final specifications for the drug substance and drug product are not expected until the end of the investigational process.
>
> It should be emphasized that the amount of information to be submitted depends upon the scope of the clinical investigation. For example, although stability data are required in all phases of the IND to demonstrate that the new drug substance and drug product are within acceptable chemical and physical limits for the planned duration of the proposed clinical investigation, if very short-term tests are proposed, the supporting stability data can be correspondingly limited."[34]

Meeting the minimum CMC regulatory requirements at each stage of clinical development is a major challenge, and really is the key to developing a successful overall corporate CMC strategy for a biopharmaceutical.

2. THE MINIMUM CMC CONTINUUM

The regulatory agencies have embraced the concept of a 'minimum CMC continuum', sometimes referred to as a graded CMC continuum, as indicated in their own guidance documents:

- "The regulations at 312.23(a)(7)(i) emphasize the graded nature of manufacturing and controls information.."[16]

- "It is not necessary to have all of the information discussed in this document available in the initial IND submission. Rather much of the information may be developed during clinical development." [29]

- "Exploratory Phase I trials for somatic cell and gene therapy products should be based on data that assure reasonable safety and rationale. Less data may be submitted to support beginning exploratory trials than may be submitted at later stages of product development, especially in the case of severe or life-threatening diseases."[35]

- "The controls used in the manufacture of APIs for use in clinical trials should be consistent with the stage of development of the drug product incorporating the API. Process and test procedures should be flexible to provide for changes as knowledge of the process increases and clinical testing of a drug product progresses from pre-clinical stages through clinical stages."[22]

Figure 4 illustrates this concept of a minimum CMC continuum. Note that there are no units expressed on the vertical axis since the actual minimum CMC values (whether it be expressed in resource commitments or in dollars expended) will depend on the specific biopharmaceutical product at hand and current regulatory agency thought on what is appropriate or required.

The minimum CMC continuum has an upward slope. Unfortunately, many biopharmaceutical companies forget this, and just view the minimum CMC requirements as a step function rather than an ongoing continuum. This causes a near panic in these companies when their biopharmaceutical rapidly shows clinical efficacy, and they then realize how far behind they are in being at the CMC level expected by the FDA.

The minimum CMC continuum also needs to be occasionally adjusted upward. CMC issues never end for a biopharmaceutical – actually, they do, but only when the biopharmaceutical product is declared dead by the company. Therefore, even if a corporate CMC strategy already exists, it needs to be periodically assessed to ensure that the plan is keeping pace with any recent new regulatory expectations for that specific type of biopharmaceutical.

**Figure 4. Illustration of the minimum CMC continuum and how the requirements
increase with advancement in clinical development**

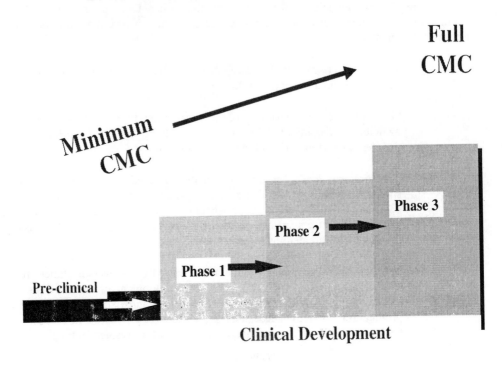

3. MINIMUM CMC REQUIREMENTS FOR CLINICAL DEVELOPMENT

The first step is to know clearly what is needed at each stage of clinical development,
from Phase 1 through Phase 3.

3.1. An Overview

The FDA has issued two guidance documents to more clearly state their expectations
for the CMC content in the various regulatory submissions required during the clinical
development stages: one document covering Phase 1 expectations[16] and one documents
covering Phase 2 and 3 requirements.[36]   Table 8 summarizes many of these CMC
expectations and provides a look through the different clinical development stages.

Table 8. **FDA's expectations for the CMC content to be submitted in the regulatory submissions that proceed through clinical development** [16, 36]

Phase 1	Phase 2	Phase 3
DRUG SUBSTANCE	**DRUG SUBSTANCE**	**DRUG SUBSTANCE**
a. Physical, Chemical, Biological Characterization A brief description of the drug substance and some evidence to support its proposed chemical structure. (It is understood that the amount of structure information will be limited in the early stage of drug development).	**a. Physical, Chemical, Biological Characterization** A more detailed description of the drug substance and updates on additional evidence to support its proposed chemical structure.	**a. Physical, Chemical, Biological Characterization** A complete description of the physical, chemical and biological characterization of the drug substance. For peptides and proteins, characterization should include data on the amino acid sequence, peptide map, post-translational modifications (e.g., glycosylation, gamma carboxylation), and secondary and tertiary structure information, if known.
b. Manufacturer The full name and street address of the manufacturer of the clinical trial drug substance.	**b. Manufacturer** The addition, deletion, or change of any manufacturer reported during Phase 1.	**b. Manufacturer** A list of all firms associated with the manufacturing and controls of the drug substance, including contract laboratories for quality control and stability testing.
c. Method of Preparation A brief description of the manufacturing process, including a list of the raw materials used. (A detailed flow diagram is suggested as the usual, most effective, presentation of this information). More information needed to assess the safety of the biopharmaceutical: cell line description and cell bank characterization	**c. Method of Preparation** An updated detailed flow diagram of the manufacturing process, including information on reagents, equipment and provisions for monitoring and controlling conditions used in each step. Any changes in cell lines, cloning, cell banks, fermentation or purification should be identified. Reprocessing procedures and pertinent controls need not be described. To the extent possible in Phase 2, sponsors should document that the manufacturing process is controlled at predetermined points and yields a product meeting tentative acceptance criteria. Although in-process controls may still be in development, information on in-process controls for monitoring adventitious agents should be provided.	**c. Method of Preparation** An updated detailed flow diagram of the manufacturing process, including information on reagents, equipment and provisions for monitoring and controlling conditions used in each step. Updates of the acceptance criteria and analytical procedures for assessing the quality of starting materials. A table listing all raw materials, including the grade of each material used, the specific identity test performed, the minimum acceptable purity level, and the step of the manufacturing process in which it is used. For special materials (e.g., proteins of animal origin), a more comprehensive list of tests, screening and acceptance criteria may be needed. A general step-by-step description of the manufacturing processes. Relevant information should indicate batch size (range),

the type of manufacturing vessel, relative ratios of reactants.

Validation of the genetic stability of the cells in production, with defined passage limits.

Controls at selected stages in the manufacturing process that ensure reaction completion, identity, and purity or proper cell growth. Tentative acceptance criteria for isolated intermediates that require control. Brief description of the analytical procedures.

Appropriate validation information available upon request.

Reprocessing procedures and pertinent controls.

d. Test Methods/Specifications

A brief description of the test methods used.

Proposed acceptance limits supported by simple analytical data of the clinical trial materials. Submission of a copy of the certificate of analysis is suggested.

Validation data and established specifications ordinarily need not be submitted at the initial stage of drug development. However, for biopharmaceuticals, preliminary specifications and additional validation data may be needed in certain circumstances to ensure safety in Phase 1.

d. Test Methods/Specifications

Identification of the characterized reference standard, if not done in Phase 1. Calibration against a recognized national or international standard if available.

Any changes in specifications.

The analytical procedure used to perform a test and to support the acceptance criteria should be indicated. A complete description of the analytical procedure and supporting validation data should be available on request.

Test results, analytical data and certificates of analysis of clinical trial material prepared since the filing of the original IND should be provided.

d. Test Methods/Specifications

Details about the synthesis and purification of the reference material. Analytical procedures used to characterize it.

Detailed listing of all the tests performed. A general description of the analytical procedures if it related to a specific USP monograph or general chapter. A complete description of non-USP analytical procedures with appropriate validation information.

Tentative acceptance criteria established for each test performed.

Impurities identified, qualified, and quantified, as appropriate. Suitable limits established based on manufacturing experience.

A summary table of updated test results, analytical data and certificates of analysis of clinical trial material prepared since the filing of the original IND.

e. Container Closure System

A brief description of the container closure system (also referred to as the packaging system)

e. Container Closure System

A brief description of any changes in the container closure system

e. Container Closure System

Detailed description of the container closure system used to transport and/or store the bulk drug substance.

f. Stability

A brief description of the stability study and the test methods used to monitor the stability of the drug substance during the toxicologic studies and the proposed clinical studies.

f. Stability

If degradation of the drug substance occurs during manufacture and storage, this should be considered when establishing acceptance criteria and monitoring quality.

f. Stability

Stress study test results, if not performed during Phase 2.

A stability protocol that includes a detailed description of the drug substance, its packaging, list of tests conducted, test methods used,

Preliminary tabular data based on representative material may be submitted. Neither detailed stability data nor the stability protocol should be submitted.

The manufacturer should propose stability-indicating analytical procedures that will detect significant changes in the quality of the drug substance.

Performance of stability stress studies with the drug substance early in drug development is encouraged.

A stability protocol should be submitted that includes a list of tests, analytical procedures, sampling time points for each of the tests and the expected duration of the stability program.

Preliminary stability data based on representative material should be provided, including all stability data for the clinical material used in the Phase 1 study.

storage conditions and duration.

A short description for each parameter being investigated demonstrating that appropriate controls and storage conditions are in place to ensure the quality of the drug substance

Tests unique to the stability program should be adequately defined and described

---

DRUG PRODUCT

a. List of Components/Specifications

A list of all components, which may include reasonable alternatives for inactive compounds, used in the manufacture of the investigational drug product, including both those compounds intended to appear in the drug product and those which may not appear, but which are used in the manufacturing process. (This list should usually be no more than one or two pages of written information).

The quality (e.g., NF, ACS) of the inactive ingredients should be cited.

For novel excipients, additional manufacturing information may be necessary.

b. Quantitative Composition

A brief summary of the quantitative composition of the investigational new drug product, including any reasonable variations that may be expected during the investigational stage. In most cases, information on component ranges is not necessary.

DRUG PRODUCT

a. List of Components/Specifications

Any change to the information specified for Phase 1.

Any change in acceptance testing for active ingredients specified for Phase 1.

Analytical procedures and acceptance criteria should be provided for noncompendial components. A brief description of the manufacture and control of these compounds or appropriate reference should be provided (e.g., DMF, NDA, BLA). Information for excipients not included in previously approved drug products should be equivalent to that submitted for new drug substances.

b. Quantitative Composition

Any change to the information specified for Phase 1.

A batch formula should be provided, if not already submitted.

DRUG PRODUCT

a. List of Components/Specifications

Updates on the information provided for Phase 2.

List of components removed during the manufacturing of the drug product.

Updates on the acceptance testing for active ingredients specified for Phase 2.

Updates on compendial excipient information specified for Phase 2. In certain cases, additional testing (e.g., functionality).

Updates on analytical procedures and acceptance criteria for noncompendial components. A full description of the manufacture and control of these compounds or appropriate reference (e.g., DMF).

b. Quantitative Composition

Updates on the information provided for Phase 2.

Quantitative batch formula.

c. Manufacturer

The full name and street address(es) of the manufacturer(s) of the clinical trial drug product..

c. Manufacturer

Updates on the information provided for Phase 1.

c. Manufacturer

List of all firms associated with the manufacturing and controls of the drug product, including contractors for stability studies, packaging, labeling and quality control release testing.

d. Method of Manufacturing

A brief, general written description of the manufacturing process, including sterilization process for sterile products. (Flow diagrams are suggested as the usual, most effective, presentations of this information).

d. Method of Manufacturing

A brief, step-by-step description of the manufacturing procedure for the unit dose should be provided. Flow diagrams should be included. Information on specific equipment used, the packaging and labeling process, and in-process controls, except for sterile products, need not to be provided.

Only safety-related information need be submitted for reprocessing procedures and controls.

For sterile products, safety updates on the manufacturing process information filed for Phase 1.

Changes to the drug product sterilization process or other changes introduced in the process to sterilize bulk drug substance or bulk drug product, components, packaging and related items. Information related to the validation of the sterilization process need not be submitted at this time.

d. Method of Manufacturing

A general, step-by-step description of the manufacturing procedure for the unit dose, including key equipment employed.

For sterile products, safety updates on the manufacturing process information filed for Phase 2.

Changes to the drug product sterilization process or other changes introduced in the process to sterilize bulk drug substance or bulk drug product, components, packaging and related items. (Information related to the validation of the sterilization process need not be submitted at this time).

e. Test Methods/Specifications

A brief description of the test methods used. For sterile products, sterility and non-pyrogenicity should be submitted.

Proposed acceptance limits. Submission of a copy of the certificate of analysis is suggested.

Validation data and established specifications ordinarily need not be submitted at the initial stage of drug development. However, for biopharmaceuticals, adequate assessment of bioactivity and preliminary specifications should be available.

Preferably, a reference standard, fully characterized, should be established. A lot can be selected as the reference standard against which initial clinical batches are

e. Test Methods/Specifications

Any changes in specifications. Any changes to the analytical procedures.

The analytical procedure used to perform a test and to support the acceptance criteria should be indicated. A complete description of the analytical procedure and supporting validation data should be available on request.

Test results, analytical data and certificates of analysis of clinical trial material prepared since the filing of the original IND should be provided.

Data updates on the degradation profile should be provided so safety assessments can be made.

e. Test Methods/Specifications

Updates on the information provided for Phase 2..

General description of the analytical procedures used, including a citation to the specific USP monograph, general chapter, or the sponsor's standard test procedure number.

Full description of the non-USP analytical procedures used, with appropriate validation information, and stated acceptance criteria.

Degradation products identified and qualified.

tested prior to release.

f. Container Closure System
Packaging procedures should be described.

f. Container Closure System
A brief description of any changes

f. Container Closure System
Updates on the information provided for Phase 2.
Name of manufacturer and supplier.
Statement on whether each component meets USP criteria.
DMF reference, if available.

g. Stability
A brief description of the stability study and the test methods used to monitor the stability of the drug product during the toxicologic studies and the proposed clinical studies, packaged in the proposed container/closure system and storage conditions. Preliminary tabular data based on representative material may be submitted. Neither detailed stability data nor the stability protocol should be submitted

g. Stability
Stability of the reconstituted solution, when applicable, should be studied and data provided.
Stress studies with the drug product should be provided.
A stability protocol should be submitted that includes a list of tests, analytical procedures, sampling time points for each of the tests and the expected duration of the stability program.
Preliminary stability data based on representative material should be provided, including all stability data for the clinical material used in the Phase 1 study.

g. Stability
Stress study test results, if not performed during Phase 2.
A stability protocol that includes a detailed description of the drug product under investigation, its packaging, list of tests to be conducted, analytical procedures, sampling time points for each of the tests, temperature/humidity conditions to be studied, and the expected duration of the accelerated and long-term stability program.
Tabulated data including the lot number, manufacturing site, and the date of manufacture of the drug product lot, and the drug substance used to manufacture the lot. Each table to contain data from only one storage condition.
Individual data points for each test, with representative chromatograms and spectra, when applicable.
Short description for each parameter being investigated in the stability program studies (i.e., stress, long-term and accelerated studies) demonstrating the appropriate controls and storage conditions used in clinical trials.
Adequate definition and description of tests unique to the stability program.
For sterile products, development of a container closure challenge tests for future stability protocols, to be considered.
Discussion of how the selected test relates to the integrity of the container.

3.2. Phase 1

For Phase 1, the IND submission must provide FDA with the CMC information and data it needs to assess the safety of the proposed Phase 1 study. It is expected that the entire submission (CMC, preclinical and clinical sections) should usually not be larger than two to three, three inch, 3-ring binders.

Although the FDA requires only brief CMC descriptions in the Phase 1 IND submission, that doesn't mean that more CMC details are not available to the biopharmaceutical manufacturer. For example, in the Phase 1 IND submission, the company needs to state whether it believes (1) the chemistry of either the drug substance or the drug product, or (2) the manufacturing of either the drug substance or the drug product, presents any signals of potential human risk; and if so, these signals of potential risks should be discussed, and the steps proposed to monitor for such risk(s) should be described, or the reason(s) why the signals should be dismissed should be discussed. For the biopharmaceutical company to accomplish this assessment, it will need more thant a 'brief' understanding of the CMC content of its product.

How important is it to meet the minimum CMC regulatory compliance concerns even at the Phase 1 stage? Take note, if the FDA considers your product to present an unreasonable health risk to the patient, they will place a clinical hold on your Phase 1 IND submission. According to their own manual of policies and procedures, a clinical hold can be placed when any of the following CMC risks arise[37]:

- A product made with unknown or impure components

- A product possessing chemical structures of known or highly likely toxicity

- A product that cannot remain chemically stable throughout the testing program proposed

- A product with an impurity profile indicative of a potential health hazard or an impurity profile insufficiently defined to assess a potential health hazard

- A poorly characterized master or working cell bank

As mentioned above, one of the three elements in the overall corporate CMC strategy is to ensure that any unique CMC concerns for a specific biopharmaceutical are adequately addressed. Because of the public concerns raised by past gene therapy trials, the FDA has required additional CMC information at the Phase 1 clinical trial stage for the gene therapy virus vector biopharmaceuticals. The FDA has highlighted certain manufacturing and QA/QC issues as critical for safety due to the unique nature of these products.[38] In fact, the FDA has indicated that if the information is not provided or they feel that the information provided is inadequate, they can put the clinical trial on hold until the information needed is addressed.[39] Table 9 presents some of the additional CMC information required to initiate Phase 1 gene therapy trials.

Table 9. Additional CMC expectations for the regulatory submissions required for gene therapy biopharmaceuticals at Phase 1 stage[38]

Provide a list of all lots of gene therapy products, cell banks (CB), and viral banks (VB), ever produced or generated in your facility for potential use in non-clinical or clinical studies of human gene therapy. Please include the date of manufacture for each, their use (e.g., non-clinical or clinical), and indicate their interrelationship, i.e., which CBs and/or VBs were used to prepare each CB, VB, or product lot

Please submit all lot release data and characterization testing for each lot of product used in clinical trials, and testing information for all master CB, working CB, master VB and/or working VB used during manufacture of your lots.

If any lots of product were produced for, but not used in, clinical studies please describe the reason they were not used.

Please provide a summary of your product manufacturing quality assurance (QA) and quality control (QC) programs. This should consist of a brief (approximately three pages) description of your system for preventing, detecting, and correcting deficiencies that may compromise product integrity or function, or may lead to the possible transmission of adventitious infectious agents. Also, identify each individual who has authority over the QA and QC programs and list their duties. Please provide the date of your last QA and QC audits of your manufacturing operations and those of contract manufacturers, vendors or other partners.

3.3. Phase 2 and 3

For advancement to Phase 2, the IND amendments must provide FDA with any updated CMC information since the Phase 1 IND submission. The phrase 'more detailed' is used at this stage of clinical development. For biopharmaceuticals, there is the specific requirement to initiate stress studies as part of the stability program.

Between the Phase 2 and the Phase 3 clinical development, there occurs the important CMC meeting referred to as the EOP2 meeting, discussed in Chapter 2. Much of the CMC information packaged in the FDA briefing document for that meeting will end up in the Phase 3 IND submission.

For advancement to Phase 3, the IND amendments must provide FDA with both updated Phase 2 CMC data and additional CMC information related to Phase 3 safety issues. The phrases 'complete' and 'detailed' are used at this stage of clinical development.

Risk management of the minimum CMC continuum requires knowing clearly what CMC is needed at the end, not just what is needed at the current stage of clinical development. One shouldn't underestimate the effort it takes to properly manage this minimum CMC continuum, and to keep it at the required level as the clinical development program advances.

4. FULL CMC REQUIREMENTS FOR DOSSIER FILING

While we can refer to a minimum CMC continuum during the clinical development stages, the requirements change to 'full and complete' CMC at the time of applying for market approval. Because of the importance of the Biologic License Application (BLA) and the New Drug Application (NDA) regulatory submission, the FDA has provided detailed guidance on the CMC content necessary for biopharmaceuticals that are recombinant proteins and monoclonal antibodies.[40]   The current format, but not the content required, for completing the CMC portion of the BLA and NDA is undergoing change. ICH is attempting to harmonize the regulatory submission formats accepted by the three regional regulatory agencies (Japan, Europe and the United States). The ICH Common Technical Document (CTD) M4 Quality Modules has been published.[41]

As shown in Figure 5, modules Q2.3 and Q3 will contain the formatted CMC information in the CTD.

Figure 5. Diagrammatic representation of the ICH Common Technical Document; obtained from FDA website: www.fda.gov/cder/guidance/4539q.pdf

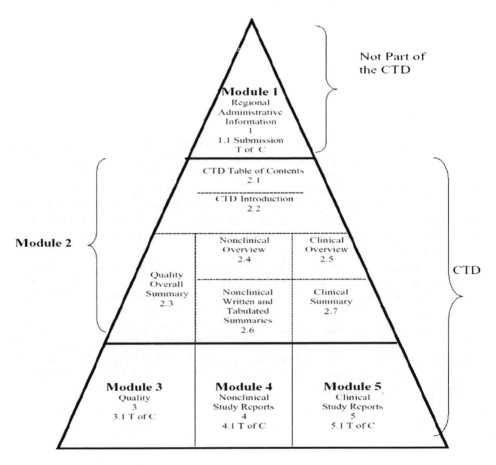

4.1. Comparison of BLA/NDA and CTD CMC Formats

The CTD describes only the CMC format required for a dossier; it does not address CMC content. On the other hand, the current FDA guidance for the BLA/NDA describe both CMC format and content. Unfortunately, the CMC formats between the CTD and BLA/NDA are different. The CMC format outlines from both the BLA/NDA and the CTD are provided in Table 10.

The FDA is releasing guidance documents to assist in the reformatting efforts to ensure that the CMC content does not get misplaced in the submitted dossiers.[42]

Currently, the CTD format is mandatory for the EMEA. For the FDA, the CTD format is still optional, but encouraged.

Table 10. Comparison of CMC: BLA/NDA versus CTD formats[40, 41]

BLA/NDA	ICH CTD M4Q
I. Drug Substance	3.2.S Drug Substance
A. Description and Characterization	3.2.S.1 General Information
1. Description	3.2.S.1.1 Nomenclature
2. Characterization/Proof of Structure	3.2.S.1.2 Structure
a. Physicochemical Characterization of Reference Standard and Qualifying Lots	3.2.S.1.3 General Properties
b. Biological Activity	3.2.S.2 Manufacture
B. Manufacturer(s)	3.2.S.2.1 Manufacturers
1. Identification	3.2.S.2.2 Description of the Manufacture Process and Process Controls
2. Floor Diagram	3.2.S.2.3 Control of Materials
3. Other Products	3.2.S.2.4 Controls of Critical Steps and Intermediates
4. Contamination Precautions	3.2.S.2.5 Process Validation and/or Evaluation
C. Method(s) of Manufacture	3.2.S.2.6 Manufacturing Process Development
1. Raw Materials and Reagents	3.2.S.3 Characterization
2. Flow Charts	3.2.S.3.1 Elucidation of Structure and Other Characteristics
3. Detailed Description	3.2.S.3.2 Impurities
a. Animal Sources	3.2.S.4 Control of Drug Substance
b. Cellular Sources	3.2.S.4.1 Specification
i. Cell Substrate/Host Cell/Expression Vector System	3.2.S.4.2 Analytical Procedures
ii. Cell Seed Lot System	3.2.S.4.3 Validation of Analytical Procedures
iii. Cell Growth and Harvesting	3.2.S.4.4 Batch Analyses
c. Purification and Downstream Processing	3.2.S.4.5 Justification of Specification
D. Process Controls	3.2.S.5 Reference Standards or Materials
1. In-Process Controls	3.2.S.6 Container Closure System
2. Process Validation	
a. Validation Studies for the Cell Growth and Harvesting Processes	

In the next chapters in this book, the CMC content required for the BLA/NDA, using either format, will be extensively discussed.

4.2. Adequate Resources Required to Compile the Full CMC Dossier

It is very important that senior management authorizes adequate resources to accomplish the task of assembling all of the required CMC content for the dossier, and that the work is initiated in a timely manner. Personnel resources need to be viewed not as person-hours but person-months to prepare the CMC dossier. Some CMC sections, such as product stability or process validation, can take longer than 6 months to generate the CMC data that then needs to be compiled and described in the dossier. Remember, if you don't have the CMC data, it is most difficult to write up that specific section in the dossier. When the FDA states that they want full CMC information provided in the dossier, they mean exactly that.

How important is it to meet the CMC regulatory compliance for the BLA/NDA dossier? Take note, if the FDA considers your CMC submission to be incomplete, they can refuse to accept your submission. According to their own manual of policies and procedures, a refusal to file can occur due to any of the following CMC deficiencies[31]:

- Incomplete description of the source material (including characterization of relevant cell banking systems) and the product

- Incomplete description of production of bulk, including description of fermentation, harvest and purification processes, process validation and testing

- Incomplete description of manufacturing steps for production through finishing, including formulation, filling, labeling and packaging (including all steps performed at outside (e.g., contract facilities)

- Omission of data demonstrating consistency of manufacture

- Absence of formulation development studies, where appropriate, or rationale for selection of formulation

- Incomplete description of lots and manufacturing process utilized for clinical studies

- Omission of data demonstrating absence or removal of adventitious agents where such agents are potentially present

- Incomplete data demonstrating equivalency to clinical trial product when significant changes in manufacturing processes or facilities have occurred

- Failure to describe changes in the manufacturing process, from material used in clinical trial to commercial production lots

- Absence of data establishing stability of the product through the dating period, and a stability protocol describing the test methods used and time intervals for product stability assessment

- Absence of the description of any of the following: (1) all manufacturing sites including contract facilities, (2) all major equipment, processes and systems, (3) validation protocols and data summaries of all major equipment, processes and systems, (4) complete flow diagrams for handling of raw materials, product, personnel air and waste, and (5) environmental monitoring data summaries (both viable and nonviable)

For many products, it can take over 2000 hours alone just to summarize all of the completed CMC reports and place them into the proper FDA format for the dossier. This task cannot be rushed.

4.3. Quality of CMC Content Present in Dossier is Critical

The quality of your CMC dossier goes a long way to leaving a positive (or negative) impression with the regulatory agency. Keep in mind that the FDA CMC reviewers have a job to do, and by the organization and attention to detail in your filed dossier, you can either make their job difficult or easy.

The following example from the FDA CMC reviewer for the submitted BLA dossier for market approval of Ontak®, recombinant denileukin diftitox, illustrates how frustrating it can become for the reviewers:

> "This review is organized according to the guidance for industry document 'For the Submission of Chemistry, Manufacturing, and Controls Information for a Therapeutic Recombinant DNA-derived Product or a Monoclonal Antibody

Product for In Vivo Use'. The submission was also organized in that manner, however, many things were cross referenced and this made it very confusing to review.

It is likely that many, if not all, analyses contained in the application will need to be repeated and the data submitted.

Since, unlike biologics, Ontak®, has not been extensively purified, Seragen has provided a section characterizing the active species in their product, in contravention to official FDA guidance. The 'specified' designation remains discretionary and the present product does not meet the requirements for this designation.

A list of raw materials has been provided, however, no certificates of analysis from the suppliers and/or manufacturer's acceptance criteria have been included.

From the CMC guidance document, a complete visual representation of the manufacturing process should be provided. No such flow chart has been included in this submission (that I could find).

An unexecuted batch record was provided. A completed (executed) representative batch record of the process of production of the drug substance should be submitted.

The descriptions of the tests are not complete enough to judge their suitability for lot release tests. A complete description of each test and the validations of the tests should have been provided."[43]

One last point, ensure that the CMC information submitted in the dossier is not only complete but also accurate. Furthermore, ensure that the CMC source data for the submission are retained. These two areas can come back later and haunt a company when the FDA inspectors are at their door.

The following example obtained from the report of an FDA inspection of the manufacturer of recombinant Factor VIII illustrates what can happen when a company cannot find the original data submitted to the FDA four years earlier in their dossier. This example also illustrates the importance of ensuring the accuracy of the information provided in the dossier:

"Failure to verify and assure the accuracy of information submitted to FDA in the license application for approval of Antihemophilic Factor (Recombinant) Sucrose Formulated (rFVIII-SF), now marketed as Kogenate FS. BLA #10332, original submission dated June 10, 1998, was approved by the agency on June 26, 2000 per Reference Number 98-0656.

Discrepancies were noted during review of quality control raw data, test records, stored test results and batch records from process validation lots when compared with information in four studies included in the BLA submission.

Quality control record discrepant information included, but was not limited to the following: tests were performed but were submitted as not performed; missing raw data, test records and stored test results; calculation errors, transcription errors, rounding errors; and failures to perform repeat testing as required after test control failures.

Recovery (%) values submitted in Tables 14, 17, 19, 22 and 24 were different than corresponding values in batch records.

Lot AZK010 original test records could not be located for Ultrafiltration Concentrate BLA Study T.02.065

Information was not available to support all production batch record data submitted."[44]

5. 'CASE-BY-CASE' CMC STRATEGY SPECIFICS

Conserving resources and trying to manage the minimum CMC continuum is a reasonable economical pathway to pursue, but it does incur risk to the biopharmaceutical company. Although obvious, it does bear restating, that unless the company achieves the minimum CMC requirements expected by the regulatory agencies at each stage of clinical trial development, the company is at tremendous disadvantage when it needs to discuss issues with the regulatory agency or prepare for the appropriate regulatory submission.

The goal of the next several chapters will be to fill in the minimum CMC continuum for biopharmaceutical products. As is indicted in Figure 6, the minimum is reasonably clear at the two ends of the development phase: the Phase 1 clinical development stage and the BLA/NDA filing stage. The tough part is figuring out how to close the CMC requirements gap between the two stages. It cannot be emphasized enough, that there is no magic formula that can be used for all biopharmaceuticals. One still has to develop their own corporate CMC strategy based on the specific product at hand, the stage of clinical trial development and current regulatory agency thought. The developed strategy is still 'case-by-case'!

Figure 6. The challenge of filling in the CMC gaps for the minimum CMC continuum for a biopharmaceutical

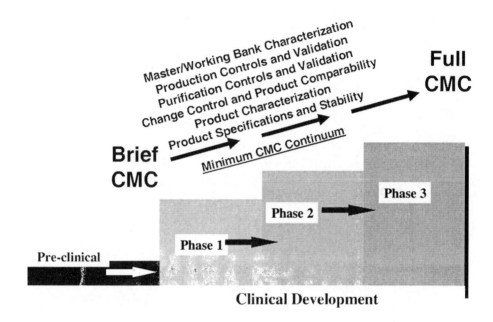

For those companies who are risk adverse, and/or with deep financial pockets, you can ignore the minimum CMC continuum recommendations provided in the following chapters and do all the CMC work that you can, as soon as you can. But for those companies that have to conserve their limited resources and are unable to do all the CMC work upfront, the CMC recommendations for both the control of the manufacturing process and the biopharmaceutical product itself provided in the following chapters should be of great help. The emphasis will be placed on 'value-added' CMC activities throughout the clinical development stages.

4

Can't Ignore cGMPs

**"Unfortunately the biologics industry has not gotten
the message about the FDA's increased
focus on GMP compliance."**

Deborah D. Ralston, Director. Office of Regional Operations
for the FDA Office of Regulatory Affairs (ORA),
speech to Food and Drug Law Institute, December 1999

1. NOT OPTIONAL

Chemistry, Manufacturing and Controls (CMC) is the body of information that
includes the manufacturing, testing, and release of product. Good manufacturing
practices are the actual requirements that need to be followed to ensure that the product
meets its intended characteristics. One cannot be CMC regulatory compliant if one is not
following good manufacturing practices.

1.1 What are 'cGMPs'?

Regulatory agencies around the world have issued performance requirements that
drug manufacturers are expected to follow. These regulations are referred to as either
good manufacturing practices (GMPs) or current good manufacturing practices (cGMPs).
The two main underlying concepts of all GMPs are:

1. A recognition that quality cannot be inspected or tested into a drug product

2.   An acceptance that each step of a manufacturing process must be controlled to maximize the likelihood that the drug product will be acceptable and safe for its intended use

Few would disagree with these two fundamental concepts; after all, they also make good business sense.

In the U.S., GMPs for drug manufacturing are requirements found in legislation and FDA regulations. In Europe, GMPs are considered part of the overall quality assurance system for drug manufacturing.

GMPs weren't formed in a vacuum as regulatory requirements. They are based on quality principles and practices that have evolved over time. Through the 1950's, the quality and safety of a product was determined mainly by testing it to determine whether it met pre-defined specifications. One wasn't really sure that the product was acceptable until all the test results were completed. From 1960's to the 1970's, the idea of building quality into the product began to take hold. By designing quality in from the very beginning, manufacturing processes can produce products that meet predetermined requirements. This philosophy is formalized in current GMPs.

GMPs can be summarized into seven elements:

- Protect the product from contamination and cross-contamination

- Prevent product mix-ups

- Know what is being done before doing it

- Document what really occurred

- Strive for consistency and control

- Have an independent group make the final decisions ("Quality Unit")

- Solve problems, learn from mistakes, monitor and continually improve

These seven elements of GMPs also make good business sense.

1.2 Three main GMP questions

Each biopharmaceutical company needs to determine what GMPs mean to their specific operations, and then make a determination of how they will meet those requirements. Most companies ready admit the need to follow GMPs, but many times they get hung up on the 'but' portion of three main questions:

1. GMPs do apply to finished drug product manufacturing; <u>but</u>, do they also apply to the manufacture of the active pharmaceutical ingredient (the biological substance)?

2. GMPs do apply to market approved drugs; <u>but</u>, do they also apply to clinical trial materials?

3. GMPs apply to chemically-synthesized drug products; <u>but</u>, are the GMPs for biopharmaceutical products any different?

It is how the biopharmaceutical company answers the 'but' portion of the question that will determine how closely they will adhere to FDA's expectations for GMP compliance.

2. GMPS FOR EVERYTHING

GMP regulations apply to finished drug product manufacturing, but they also apply to the manufacture of the biological substance. GMPs apply to market products, but they also apply to clinical trial materials. Furthermore, additional GMP regulations apply to the biopharmaceutical products.

2.1. For Finished Drug Products

The best known cGMP regulations are those contained in the FDA Code of Federal Regulations (CFR) Title 21 Part 211.[45] These regulations apply to all finished drug products whether of chemical origin or a biopharmaceutical. The finished drug product is the formulated active pharmaceutical ingredient (API) in a container ready for distribution for human use.

These cGMPs are FDA law, not merely recommendations or guidelines or best practices or desired systems. Drugs not manufactured in accordance with these requirements are considered adulterated under Section 501(a)(2)(B) of the Federal Food, Drug and Cosmetic Act:

> "A drug or device has be deemed to be adulterated – (1) if it consists in whole or in part of any filthy, putrid, or decomposed substance; or (2)(A) if it has been prepared, packed, or held under insanitary conditions whereby it may have been contaminated with filth, or whereby it may have been rendered injurious to health; or (2)(B) if it is a drug and the methods used in, or the facilities or controls used for, its manufacture, processing, packing, or holding do not conform to or are not operated or administered in conformity with current good manufacturing practices …."[46]

Because of this legal status, the U.S. Justice Department, working with the FDA, has powers to ensure compliance by seizing product and levying fines. Therefore, a company must take the cGMP regulations seriously.

The purpose for these regulations is to ensure that the finished drug products are consistently manufactured and controlled to the quality standards appropriate for their intended use and as required by the product specifications and regulatory agencies. Some of the basic operating principles of these cGMPs are:

- All manufacturing processes are to be clearly defined, systematically reviewed in the light of experience and shown to be capable of consistently manufacturing products of the required quality

- Critical steps of the manufacturing process and significant changes to the process are to be validated

- All necessary facilities are to be provided, including appropriately qualified and trained personnel, adequate premises and space, suitable equipment and services, correct materials and components, approved procedures and instructions, and suitable storage and transport

- Instructions and procedures are to be written in an instructional form in clear and unambiguous language

- Records are to be made during manufacture which demonstrate that all the steps required by the defined procedures and instructions were in fact taken; and that the quantity and quality of the product was as expected

Table 11 presents further details of the U.S. cGMP regulations for finished drug products. It is important to keep in mind that the cGMP requirements are really only <u>minimum</u> requirements (not maximum as some suppose), as indicated in 21 CFR 211.1 (a):

"The regulations in this part contain the minimum current good manufacturing practice for preparation of drug products for administration to humans or animals."[45]

The expectation is that drug manufacturers will meet at least the stated minimum, but preferably operate above this threshold. This is especially important since cGMPs also are subject to change or ongoing interpretation by the regulatory authorities (the 'c' in cGMP actually is for 'current', not 'creeping upward' as some believe). For example, in 21 CFR Part 211, there is currently no specific mention of the requirement to carry out an internal audit program nor specific mention of the requirement to have completed process and cleaning validation; yet, these are clear expectations of the FDA today. This is one of the reasons that companies need to consider doing more than just what is stated in the written regulation.

Table 11. Summary of cGMP requirements for drug products, including biopharmaceuticals[45]

Organization and Personnel	There shall be a quality control unit that shall have the responsibility and authority to approve or reject all components, drug product containers, closures, in-process materials, packaging material, labelling, and drug products, and the authority to review production records to assure that no errors have occurred or, if errors have occurred, that they have been fully investigated
	The quality control unit shall be responsible for approving or rejecting drug products manufactured, processed, packed, or held under contract by another company
	Each person engaged in the manufacture, processing, packing, or holding of a drug product shall have education, training, and experience, or any combination thereof, to enable that person to perform the assigned functions
	There shall be an adequate number of qualified personnel to perform and supervise the manufacture, processing, packing, or holding of each drug product
Buildings and Facilities	Any building or buildings used in the manufacture shall be of suitable size, construction and location to facilitate cleaning, maintenance, and proper operations
	Operations shall be performed within specifically defined areas of adequate size
	Adequate lighting and ventilation shall be provided in all areas
	Sewage, trash, and other refuse in and from the building and immediate premises shall be disposed of in a safe and sanitary manner
	Any building used in the manufacture, processing, packing, or holding of a drug product shall be maintained in a good state of repair
Equipment	Equipment used in the manufacture, processing, packing, or holding of a drug product shall be of appropriate design, adequate size, and suitably located to facilitate operations for its intended use and for its cleaning and maintenance
	Equipment shall be constructed so that surfaces that contact components, in-process materials, or drug products shall not be reactive, additive, or absorptive so as to alter the safety, identity, strength, quality, or purity of the drug product beyond the official or other established requirements
	Equipment and utensils shall be cleaned, maintained, and sanitized at appropriate intervals to prevent malfunctions or contamination that would alter the safety, identity, strength, quality, or purity of the drug product beyond the official or other established requirements
Control of Components	There shall be written procedures describing in sufficient detail the receipt, identification, storage, handling, sampling, testing, and approval or rejection of components and drug product containers and closures; such written procedures shall be followed
	Each lot of components, drug product containers, and closures shall be withheld from use until the lot has been sampled, tested, or examined, as appropriate, and released for use by the quality control unit
	Rejected components, drug product containers, and closures shall be identified and controlled under a quarantine system designed to prevent their use in manufacturing

Production and Process Controls	There shall be written procedures for production and process control designed to assure that the drug products have the identity, strength, quality, and purity they purport or are represented to possess
	Actual yields and percentages of theoretical yield shall be determined at the conclusion of each appropriate phase of manufacturing, processing, packaging, or holding of the drug product
	To assure batch uniformity and integrity of drug products, written procedures shall be established and followed that describe the in-process controls, and tests, or examinations to be conducted on appropriate samples of in-process materials of each batch
	When appropriate, time limits for the completion of each phase of production shall be established to assure the quality of the drug product
Laboratory Controls	The establishment of any specifications, standards, sampling plans, test procedures, or other laboratory control mechanisms required by this subpart, including any change in such specifications, standards, sampling plans, test procedures, or other laboratory control mechanisms, shall be drafted by the appropriate organizational unit and reviewed and approved by the quality control unit
	Laboratory controls shall include the establishment of scientifically sound and appropriate specifications, standards, sampling plans, and test procedures designed to assure that components, drug product containers, closures, in-process materials, labelling, and drug products conform to appropriate standards of identity, strength, quality, and purity
	For each batch of drug product, there shall be appropriate laboratory determination of satisfactory conformance to final specifications for the drug product, including the identity and strength of each active ingredient, prior to release
	There shall be a written testing program designed to assess the stability characteristics of drug products
	For each batch of drug product purporting to be sterile and/or pyrogen-free, there shall be appropriate laboratory testing to determine conformance to such requirements
Records and Reports	Any production, control, or distribution record that is required to be maintained in compliance with this part and is specifically associated with a batch of a drug product shall be retained for at least 1 year after the expiration date of the batch
	To ensure uniformity from batch to batch, master production and control records for each drug product, including each batch size thereof, shall be prepared, dated, and signed (full signature, handwritten) by one person and independently checked, dated, and signed by a second person.
	Batch production and control records shall be prepared for each batch of drug product produced and shall include complete information relating to the production and control of each batch
	Laboratory records shall include complete data derived from all tests necessary to assure compliance with established specifications and standards, including examinations and assays

Depending upon the region(s) of the world the biopharmaceutical company intends to operate, there may be additional GMP regulations to follow (e.g.., the World Health Organization, Canada, Australia, etc. each have their own GMP requirements). The GMPs for the European Union[47] cover basically the same points as 21 CFR 211.

2.2. Required for APIs Also

GMP compliance is typically thought of for finished drug products only, but it also applies to the manufacture of active pharmaceutical ingredients (APIs). The API includes all of those manufacturing steps involved in the production and purification of the active ingredient. It needs to be stressed that APIs not manufactured in accordance with GMP are considered adulterated under Section 501(a)(2)(B) of the Federal Food, Drug and Cosmetic Act, since no distinction in the act is made between the API and the finished drug product.

Regulatory agencies have accepted the International Conference on Harmonization (ICH) Q7A document which provides guidance on GMPs for APIs.[22] Although guidance documents are not the same as regulations, they do provide both industry and the regulatory agencies with an interpretation of the regulations. ICH Q7A has something for almost every market approved (regulated) active pharmaceutical ingredient (API). All APIs of chemical origin are covered by this guidance. Most biopharmaceutical APIs are covered by this guidance and that includes all recombinant protein products manufactured using mammalian, plant, insect or microbial cells and transgenic animals. Although not clearly stated, transgenic plants would also fall under the umbrella of this document. However, specifically excluded by this guidance document are the recombinant vaccines and the gene therapy biopharmaceuticals. Table 12 presents further details of the GMP requirements for these regulated APIs. The GMP regulations for the APIs are similar to the GMP regulations for the finished drug products.

Table 12. Summary of GMP requirements for APIs, including biopharmaceuticals[22]

Quality Management	Quality should be the responsibility of all persons involved in manufacturing
	Each manufacturer should establish, document, and implement an effective system for managing quality that involves the active participation of management and appropriate manufacturing personnel
	There should be a Quality Unit (s) that fulfills both QA and QC responsibilities and it should be independent on production
	All quality-related activities should be recorded at the time they are performed

	Any deviation from established procedures should be documented and explained; critical deviations should be investigated, and the investigation and its conclusions should be documented
	Procedures should exist for notifying responsible management in a timely manner of regulatory inspections, serious GMP deficiencies, product defects, quality related complaints, recalls and regulatory actions
Personnel	There should be an adequate number of personnel qualified by appropriate education, training, and/or experience to perform and supervise manufacture
	Responsibilities are to be specified in writing
	Training should be regularly conducted by qualified individuals and should cover, at a minimum, the particular operations that the employee performs and GMP as it relates to the employees functions
	Personnel should practice good sanitation and health habits
	Personnel should wear clean clothing suitable for the manufacturing activity with which they are involved and this clothing should be changed, when appropriate
Buildings and Facilities	Buildings and facilities should be located, designed, and constructed to facilitate cleaning, maintenance, and operations
	Facilities should be designed to minimize potential contamination
	There should be adequate space for the orderly placement of equipment and materials to prevent mix-ups and contamination
	All utilities that could affect product quality (e.g., steam, gas, compressed air, heating, ventilation, and air conditioning) should be qualified and appropriately monitored and action should be taken when limits are exceeded
	If air is recirculated to production areas, appropriate measures should be taken to control risks of contamination and cross-contamination
	Buildings should be properly maintained and repaired and kept in a clean condition
	Appropriate measures should be established and implemented to prevent cross-contamination from personnel and materials from one dedicated area to another
	Adequate lighting should be provided
Process Equipment	Equipment should be of appropriate design and adequate size, and suitably located for its intended use, cleaning, sanitation and maintenance
	Production equipment should only be used within its qualified operating range
	Schedules and procedures (including assignment of responsibility) should be established for the preventative maintenance of equipment
Documentation and Records	All documents related to the manufacture should be prepared, reviewed, approved, and distributed according to written procedures

	Written procedures should be established and followed for the review and approval of batch production and laboratory control records
Materials Management	There should be written procedures describing the receipt, identification, quarantine, storage, handling, sampling, testing, and approval or rejection of materials
	Materials should be handled and stored in a manner to prevent degradation, contamination, and cross-contamination
Production and In -Process Controls	Written procedures should be established to monitor the progress and control the performance of processing steps that cause variability in the quality characteristics
	In-process controls and their acceptance criteria should be defined based on the information gained during the developmental stage or from historical data
	Any deviation should be documented and explained; any critical deviation should be investigated
Laboratory Controls	The independent quality unit should have at its disposal adequate laboratory facilities
	There should be documented procedures describing sampling, testing, approval, or rejection of materials and recording and storage of laboratory data
	A documented, on-going testing program should be established to monitor the stability characteristics of APIs; the test procedures used in stability testing should be validated and be stability indicating
Validation	The company's overall policy, intentions, and approach to validation, including the validation of production processes, cleaning procedures, analytical methods, in-process control test procedures, computerized systems, and persons responsible for design, review, approval, and documentation of each validation phase, should be documented
Change Control	A formal change control system should be established to evaluate all changes that could affect the production and control of the intermediate or API
	Any proposals for GMP relevant changes should be drafted, reviewed, approved by the appropriate organizational units and reviewed and approved by the quality unit
Con tract Manufacturers	Companies should evaluate any contractors (including laboratories) to ensure GMP compliance of the specific operations occurring at the contractor sites
	There should be a written and approved contract or formal agreement between a company and its contractors that defines in detail the GMP responsibilities, including the quality measures, of each party

2.3. Extra GMPs for Biopharmaceuticals

Are there any differences in the GMP regulations between a chemically-synthesized product and a biopharmaceutical product? Yes. We have already seen that biopharmaceuticals are different than the chemically-synthesized pharmaceuticals; therefore, it is not surprising that there would be additional GMP requirements for biopharmaceutical APIs.

ICH Q7A Section 18 lays out the extra GMPs for biopharmaceuticals. These additional regulations are presented in Table 13.

Table 13. Summary of additional GMP requirements for biopharmaceutical APIs[22]

General	In general, the GMP principles in the other sections of this document apply
	In general, the degree of control for biotechnological processes used to produce proteins and polypeptides is greater than that for classical fermentation processes
	Appropriate controls should be established at all stages of manufacturing to ensure intermediate and/or API quality
	Appropriate equipment and environmental controls should be used to minimize the risk of contamination; the acceptance criteria fir determining environmental quality and the frequency of monitoring should depend on the step in production and the production conditions (open, closed or contained systems)
	Removal of media components, host cell proteins, other process-related impurities, product-related impurities and contaminants should be demonstrated
Cell Bank Maintenance and Record Keeping	Access to cell banks should be limited to authorized personnel
	Cell banks should be maintained under storage conditions designed to maintain viability and prevent contamination
	Records of the use of the vials from the cell banks and storage conditions should be maintained
	Cell banks should be periodically monitored to determine suitability for use
Cell Culture/ Fermentation	Where cell substrates, media, buffers and gases are to be added under aseptic conditions, closed or contained systems should be used where possible
	For manipulations using open vessels, there should be controls and procedures in place to minimize the risk of contamination
	Where the quality of the API can be affected by microbial contamination, manipulations using open vessels should be performed in a biosafety cabinet or similarly controlled environment
	Personnel should be appropriately gowned and take special precautions handling the cultures
	Critical operating parameters (for example, temperature, pH, agitation rates, addition

of gases, pressure) should be monitored for consistency with the established process

Cell growth, viability and ,where appropriate, productivity should also be monitored

Cell culture equipment should be cleaned and sterilized after use

Culture media should be sterilized before use, when necessary to protect the API

Appropriate procedures should be in place to detect contamination and determine the course of action to be taken; procedures should be available to determine the impact of the contamination on the product and to decontaminate the equipment and return it to a condition to be used in subsequent batches

Foreign organisms observed during fermentation processes should be identified, as appropriate, and the effect of their presence on product quality should be assessed

Records of contamination events should be maintained

Shared (multi-product) equipment may warrant additional testing after cleaning between product campaigns to minimize the risk of cross-contamination

Harvesting, Isolation and Purification

Harvesting steps, either to remove cells or cellular components or to collect cellular components after disruption, should be performed in equipment and areas designed to minimize the risk of contamination

Harvest and purification procedures that remove or inactive the producing organism, cellular debris and media components (while minimizing degradation, contamination, and loss of quality) should be adequate to ensure that the intermediate or API is recovered with consistent quality

All equipment should be properly cleaned and, as appropriate, sanitized after use; multiple successive batching without cleaning can be used if intermediate or API quality is not compromised

If open systems are used, purification should be performed under environmental conditions appropriate for the preservation of product quality

Additional controls, such as the use of dedicated chromatography resins or additional testing, may be appropriate if equipment is to be used for multiple products

Viral Removal/ Inactivation Steps

Viral removal and viral inactivation steps are critical processing steps for some processes and should be performed within their validated parameters

Appropriate precautions should be taken to prevent viral contamination from previral to postviral removal/inactivation steps; therefore, open processing should be performed in areas that are separate from other processing activities and have separate air handling units

The same equipment is not normally used for different purification steps; however, if the same equipment is to be used, the equipment should be appropriately cleaned and sanitized before reuse

Appropriate precautions should be taken to prevent potential virus carry-over (e.g., through equipment or environment) from previous steps

The GMPs for the European Union listed in the EudraLex Volume 4 Annex 2 also contain some extra regulations specific to biopharmaceuticals:

"All personnel (including those concerned with cleaning, maintenance or quality control) employed in areas where biological medicinal products are manufactured should receive additional training specific to the products manufactured and to their work. Personnel should be given relevant information and training in hygiene and microbiology.

In the course of a working day, personnel should not pass from areas where exposure to live organisms or animals is possible to areas where other products or different organisms are handled. If such passage is unavoidable, clearly defined decontamination measures, including change of clothing and shoes and, where necessary, showering should be followed by staff involved in any such production.

The degree of environmental control of particulate and microbial contamination of the production premises should be adapted to the product and the production step, bearing in mind the level of contamination of the starting materials and the risk to the finished product.

The risk of cross-contamination between biological medicinal products, especially during those stages of the manufacturing process in which live organisms are used, may require additional precautions with respect to facilities and equipment, such as the use of dedicated facilities and equipment, production on a campaign basis and the use of closed systems. The nature of the product as well as the equipment used will determine the level of segregation needed to avoid cross-contamination.

Specifications for biological starting materials may need additional documentation on the source, origin, method of manufacture and controls applied, particularly microbiological controls." [48]

These extra GMP regulations for biopharmaceuticals are derived from the way in which these products are produced, controlled and administered. It's the use of living recombinant hosts, the product themselves and the impact of the manufacturing process on the product, which results in these extra precautions.

3. WHERE IN THE MANUFACTURING PROCESS SHOULD GMP BEGIN?

GMPs must be applied at all manufacturing process steps for the finished drug product. But, as we look at the API manufacturing process steps, where in the process should GMP start? For drugs of chemical origin, it is well known that the introduction of the starting material is the place where GMP begins and must continue on downstream through the manufacturing process. But where is the starting point for a biopharmaceutical manufacturing process? Is it during the genetic engineering steps when DNA is manipulated, during the preparation of cell/seed banks, or maybe not until during cell culture operations or purification?

ICH Q7A Section 1 states clearly that GMP for a biopharmaceutical manufacturing process begins during the maintenance of the working cell bank, and that it continues on .downstream through the manufacturing process, as described in Table 14.

Table 14. Start of GMP in the manufacturing process[22]

Type of Manufacturing	Start of GMP in Manufacturing Process
Chemical Manufacturing	Introduction of the starting material into the process
Animal Sourced	Introduction of the starting material into the process
Plant Sourced	Introduction of the starting material into the process
Classical Fermentation	Introduction of the cells into fermentation
Biopharmaceuticals	Maintenance of the working cell bank

But does that mean that GMP should not be applied before the maintenance of the working bank step, say for example, the preparation of the master bank? The answer is absolutely 'No'! ICH Q7A Section 1 also states:

> "Note that cell substrates (mammalian, plant, insect or microbial cells, tissue or animal sources including transgenic animals) and early process steps may be subject to GMP but are not covered by this guidance.
>
> While this guidance starts at the cell culture/fermentation step, prior steps (e.g., cell banking) should be performed under appropriate process controls"[22]

Just because ICH Q7A does not cover certain process steps, does not mean that GMP should not be applied. As we will see in the next chapter on master and working banks, the application of GMPs, or at least the application of appropriate process controls, are important in these early process steps to provide sufficient documentation and traceability, and to adequately protect the product.

4. WHEN DURING CLINICAL DEVELOPMENT SHOULD GMP BEGIN?

It is clear that GMPs must be in place for the manufacture of market approved drug products. However, GMP compliance also applies to the clinical development stages. GMP is necessary even at Phase 1 clinical trials! Now that that statement is out of the way, let's discuss why. Again think of GMP as good business practice or good common sense, and the need to protect the patient becomes obvious, whether that patient has entered an early stage clinical trial or a later stage clinical trial, and whether that patient has a terminal illness or not. If you were the recipient of your biopharmaceutical, what manufacturing controls would you want in place for the product?

Regulatory agencies clearly expect compliance with GMPs during the clinical development stage for both API manufacturing and finished drug product manufacturing.

4.1. API Clinical Trial Materials

International agreement on what GMPs are necessary for API clinical trial material (CTM) manufacturing is contained in ICH Q7A Section 19.[22] GMP compliance is clearly stated in this document, but it is also clearly stated, that the level of requirements is reduced for clinical trial materials compared to that required for market approved (regulated) products:

> "Not all controls in the previous sections of this guidance are appropriate for the manufacture of a new API for investigational use during its development.
>
> The controls used in the manufacture of APIs for use in clinical trials should be consistent with the stage of development of the drug product incorporating the API. Process and test procedures should be flexible to provide for changes as knowledge of the process increases and clinical testing of a drug product progresses from pre-clinical stages through clinical stages. Once drug development reaches the stage where the API is produced for use in drug products intended for clinical trials, manufacturers should ensure that APIs are manufactured in suitable facilities using appropriate production and control procedures to ensure the quality of the API."[22]

Table 15 presents further details of the GMP requirements for clinical trial material APIs, including biopharmaceuticals. Note that the levels required of system development, of documentation, and of validation are reduced compared to that necessary for market approved products:

- Requirement to adequately record all changes during clinical development versus the need for a formal change control system at market approval

- The need for appropriate documentation during clinical development versus the need for a formal document control system at market approval

- Process validation and test method validation not required during clinical development versus mandatory at market approval

But also note that product safety is not compromised for the manufacture of clinical trial materials (e.g., procedures for use of facilities should ensure that materials are handled in a manner that minimizes the risk of contamination and cross-contamination).

Table 15. Summary of GMP requirements for all clinical trial material APIs, including biopharmaceuticals[22]

General	Not all controls in the previous sections of this guidance are appropriate for the manufacture of a new API for investigational use during its development
	The controls used in the manufacture of APIs for use in clinical trials should be consistent with the stage of development of the drug product incorporating the API
Quality	Appropriate GMP concepts should be applied in the production of APIs for use in clinical trials with a suitable mechanism for approval of each batch
	A quality unit(s) independent from production should be established for the approval or rejection of each batch of API for use in clinical trials
	Some of the testing functions commonly performed by the quality unit can be performed within other organizational units
	Quality measures should include a system for testing of raw materials, packaging materials, intermediates, and APIs
	Process and quality problems should be evaluated
Equipment and Facilities	During all phases of clinical development, including the use of small-scale facilities or laboratories to manufacture batches of APIs for use in clinical trials, procedures should be in place to ensure that equipment is calibrated, clean and suitable for its intended use
	Procedures for the use of facilities should ensure that materials are handled in a manner that minimizes the risk of contamination and cross-contamination
Control of Raw Materials	Raw materials used in production of APIs for use in clinical trials should be evaluated by testing, or received with a supplier's analysis and subjected to identify testing
	In some instances, the suitability of a raw material can be determined before use based on acceptability in small-scale reactions (i.e., use testing) rather than on analytical testing alone

Production	Production should be documented in laboratory notebooks, batch records, or other appropriate means
	These documents should include information on the use of production materials, equipment, processing and scientific observations
	Investigations into yield variations are not expected
Validation	Process validation for the production of APIs for use in clinical trials is normally inappropriate, where a single API batch is produced or where process changes during API development make batch replication difficult or inexact
	The combination of controls, calibration, and, where appropriate, equipment qualification ensures API quality during this development phase
	Process validation should be conducted when batches are produced for commercial use, even when such batches are produced on a pilot or small scale
Changes	Changes are expected during development, as knowledge is gained and the production is scaled up
	Every change in the production, specifications, or test procedures should be adequately recorded
Laboratory Controls	While analytical methods performed to evaluate a batch of API may not yet be validated, they should be scientifically sound
	A system for retaining reserve samples of all batches should be in place; this system should ensure that a sufficient quantity of each reserve sample is retained for an appropriate length of time after approval, termination, or discontinuation of an application
	For new APIs, expiry dating does not normally apply
Documentation	A system should be in place to ensure that information gained during the development and the manufacture of APIs is documented and available
	The development and implementation of the analytical methods used to support the release of a batch of API should be appropriately documented
	A system for retaining production and control records and documents should be used, and the system should ensure that they are retained for an appropriate length of time after the approval, termination or discontinuation of an application

4.2. Drug Product Clinical Trial Materials

But what are the GMP expectations for the manufacture of clinical trial material (CTM) finished drug products? The FDA doesn't formally distinguish between GMP requirements for market approved and clinical drug products, and this causes concern in the industry. However, the FDA recognizes that during clinical development, a company will be developing their control systems, improving their documentation and completing their validation.

Europe on the other hand has taken the bold step of actually releasing a proposed set of GMPs for medicinal preparations to be used in clinical studies.[50] They have done an excellent job of outlining a philosophy of providing manufacturing flexibility during this stage but yet also providing adequate product safety:

> "The application of GMP to the manufacture of investigational medicinal products is intended to ensure that trial subjects are not placed at risk, and that the results of clinical trials are unaffected by, inadequate safety, quality or efficacy arising from unsatisfactory manufacture. Equally it is intended to ensure that there is consistency between batches of the same investigational medicinal product used in the same or different clinical trials or that changes during the development of an investigational medicinal product are adequately documented and justified.
>
> The production of investigational medicinal products involves added complexity in comparison to marketed products by virtue of the lack of fixed routines, variety of clinical trial designs, consequent packaging designs, the need, often, for randomisation and blinding and increased risk of product cross-contamination and mix up. Furthermore there may be incomplete knowledge of the potency and toxicity of the product and a lack of full process validation.
>
> These challenges require personnel with a thorough understanding of, and training in, the application of GMP to investigational medicinal products.
>
> This increased complexity in manufacturing operations requires a highly effective Quality system." [49]

5. CONSEQUENCES OF NOT FOLLOWING GMPS

Although six key words can define the essence of GMP – appropriate, adequate, controlled, approved, documented, and traceable – the challenge comes in the implementation of these key words into our manufacturing processes and testing of the biopharmaceutical products. The consequence of not meeting GMP is the risk that our product does not meet the requirements for identity, strength, quality, and purity that it purports or is represented to possess. The overall corporate CMC strategy must factor in these GMP requirements of the regulatory agencies.

5.1. Issues with Market Approved Biopharmaceuticals

GMPs must be followed. Table 16, which lists two Warning Letters issued by the FDA, presents examples of some compliance issues that FDA inspectors have found when inspecting biopharmaceutical companies.

Table 16. Some cGMP compliance problems reported by the FDA during inspections of biopharmaceutical manufacturers[50, 51]

Warning Letter to Bayer Corporation 2001

Failure to thoroughly investigate any unexplained discrepancy or the failure of a batch or any of its components to meet any of its specifications [21 CFR 211.192], as follows:

(a) Investigations into pyrogen failures were limited and did not include any prior processes
(b) Investigations into container integrity failures failed to address other finished product lots which have used the implicated lots of glassware
(c) Ten lots of ultrafiltered clarified tissue culture fluid material that exceeded the bioburden limit were released for further processing without determining an assignable cause for the increased microbial load
(d) Investigations into microbial excursion results for the water for injection loops are incomplete in that there is no documentation of the recommendations for further investigations, corrective actions and follow-up

Failure to establish and follow written procedures for the cleaning and maintenance of equipment, including utensils, used in the manufacture, processing, packing, or holding of a drug product [21 CFR 211.67(b)] in that:

(a) The cleaning of the ultrafiltration/diafiltration (UF/DF) unit used in manufacture has not been adequately validated.
(b) The UF/DF filter cartridges have no established maximum number of uses.

Failure to establish and follow written procedures applicable to the quality control unit [21 CFR 211.22(d)]. For example:

(a) There is no written procedure in place to track all batch record corrections, explanations, or required deviations from manufacturing.
(b) The SOP entitled "QA Release Department Review of Batch Production Records and Test Results," is inadequate in that it allows manufacturing supervisors to make changes to batch records without the knowledge or consultation of the manufacturing employees that were involved in the discrepancies.

Warning Letter to Genentech, Inc. 2000

Failure to obtain approval from the quality control unit prior to reprocessing [21 CFR 211.115(b)] in that, there is no documentation that the quality control unit was notified prior to the re-filtration of Activase lot #L9042A. On April 10, 2000, during set up for filling, manufacturing detected a leak in the connection during set up for priming. On April 11, 2000, the bulk was re-filtered by manufacturing and filled as Activase lot #L9046A.

Failure to follow or maintain written procedures and to record and justify any deviation from written procedures for production and process control designed to assure that the drug products have the identity, strength, quality and purity they purport or are represented to possess [21 CFR 211.100]. On January 15, 2000, during the manufacture of Pulmozyme bulk lot, an expired concentrated bulk was used. There is no indication that the impact to material, which was held for an extended period of time, was evaluated.

Failure to submit a supplement and obtain approval prior to distribution of a product following any change in the product, production process, quality controls, equipment, facilities, or responsible personnel that has a substantial potential to have an adverse effect on the identity, strength, quality, purity, and potency of the product as they may relate to the safety or effectiveness of the product [21 CFR 601.12(b)] in that, two Pulmozyme thawed bulk lots, G90536/PRK11930 and G90536/K14225, were re-filtered after brown foreign material and brown particulates were observed. Final vial lot K9721A was released by Quality Assurance and subsequently distributed.

It should be noted that European inspectors encounter similar GMP compliance problems during their inspections of drug manufacturers. Since Europe does not have a freedom of information act like out FDA, it is not easy to follow the GMP concerns that they are observing. However, from presentations at conferences by their inspectors, the following are three of the major GMP compliance issues that they have reported[52]:

- Batches being released to the market intentionally even after a non-compliance with the marketing license has been identified

- No quality system established to ensure that the master batch record is in accordance with the quality dossier submitted and approved by the regulatory agency

- No system established by the manufacturer to identify a need for change in the marketing license nor a system that required the manufacturer to wait for approval of a change before release of the product

Another viewpoint of what are some of the major GMP compliance issues with products already on the market, can be gleaned from the FDA summaries of Biological Product Deviation Reports (BPDRs), previously known as Error and Accident Reports. If a biopharmaceutical has been released into market distribution and a major quality or safety defect is discovered, the manufacturer has 45 days to from date of discovery to notify CBER by submitting the BPDR (21, CFR 600.14)[53, 54] It should be noted that if the biopharmaceutical is regulated by CDER, then a NDA Field Alert (21 CFR 314.81) should be filed with within 3 days to CDER instead.[55] Each quarter CBER publishes a summary of the BPDRs that it has received. Table 17 presents the CBER summary for FY02.

Table 17. BPDRs reported to CBER in FY02 for non-blood therapeutic products[56]

BPDR Filing Category	Total Number	Reasons for the BPDR
Incoming Material Specifications	2	Specifications for container not met
		Raw material found to be unsuitable
Process Controls	7	Manufactured using incorrect parameters
		Incorrect procedure performed (3)
		Environmental monitoring excursions (2)
		Process equipment not qualified
Testing	4	Purity measurement incorrectly performed (2)
		Sterility test incorrectly performed
		Identity test not performed
Labeling	1	Expiration date improperly extended
Product Specifications	6	Product specifications not met (2)
		Preservative content spec not met
		Potency failed on stability
		Purity failed on stability
		Defective administration set
Quality Control & Distribution	3	Product distributed inappropriately
		Improper shipping and storage
		Product shipped at incorrect temperature
Miscellaneous	0	

The EMEA has recently established a similar program for handling of reports of suspected quality defects in medicinal products.[57]

5.2. Issues During Clinical Development

But do GMPs really matter during the clinical development stages? After all, if the regulatory agencies don't inspect these products, why should a company expend the resources to meet the clinical cGMP requirements? This is the 'if we won't get caught, we can do what we want' mentality.

Does the FDA ever inspect clinical trial materials for cGMP compliance violations? Yes! They have the legal right to do so according to 21 CFR 312.44(b):

> "FDA may terminate an IND if it finds that the methods, facilities, and controls are inadequate to establish and maintain appropriate standards of identity, strength, quality, and purity as needed for patient safety."[58]

Although the FDA currently does not have the resources to inspect all clinical trial materials, they will carry out an inspection if they have concerns about patient safety. Several incidents where this occurred are presented in Table 18.

Table 18. FDA cGMP inspections of clinical trial materials[59, 60]

Warning Letter to AVAX Technologies, 2001

"The purpose of the investigation was to review AVAX's activities as the sponsor and manufacturer of investigational autologous melanoma tumor vaccines."

"Your firm routinely manufactured tumor vaccines from tumor source materials that were previously shown to be non-sterile. You manufactured additional lots of vaccine from the original tumor source material even though the first lots of vaccine were proven to be contaminated. Your practice of releasing contaminated vaccines repeatedly exposed subjects to increased risk."

"You failed to take appropriate corrective action to develop improved aseptic sampling procedures and to institute new training procedures to prevent the contamination of tumor source materials provided to AVAX."

Letter to Cell Therapy Research Foundation, 2001

"Investigational New Drug Application (IND) for cultured allogeneic myoblasts, and cyclosporin (Sandoz)."
"The Agency's inspection of your establishment from July 12 – September 15, 1999, disclosed serious deficiencies in quality control and quality assurance of product manufacturing. During the inspection, the Agency found that the frozen allogeneic myoblasts in CTRF's inventory were not manufactured in conformity with current Good Manufacturing Practice (cGMP) regulations."

"The inspection findings included the improper use of reagents that were labelled as not for human use, the lack of documentation of required testing, and the clinical use of cells of questionable sterility or viability."

It should be noted that Europe has taken a more ambitious approach to controlling clinical trial materials. They have issued a clinical trials directive (2001/20/EC), that when made effective in the near future, will require clinical trial materials to be manufactured in a GMP-compliant facility which has been GMP inspected.[61]

5.3. How to Avoid GMP Difficulties with the FDA

To avoid difficulties with the FDA, know what GMP concerns the inspectors are focusing upon and then avoid the system problems that they consider serious. This is really not so hard, since the FDA publishes both how it trains its inspectors and a list of what they consider major system failures. Table 19 presents the examples of major system failures listed in the FDA Compliance Program Guidance Manual.[62] Avoid these! Manufacturers who choose to wait until the FDA inspectors find GMP violations, rather than policing themselves, will find that they have made a poor and costly decision.

Table 19. Significant cGMP deficiencies and system failures[62]

Quality System	1. Pattern of failure to review/approve procedures
	2. Pattern of failure to document execution of operations as required
	3. Pattern of failure to review documentation
	4. Pattern of failure to conduct investigations and resolve discrepancies/failures/deviations/complaints
	5. Pattern of failure to assess other systems to assure compliance with GMPs and SOPs
Facilities and Equipment	1. Contamination with filth, objectionable microorganisms, toxic chemicals, other drug chemicals, or a reasonable potential for contamination, with demonstrated avenues of contamination, such as airborne or through unclean equipment
	2. Pattern of failure to validate cleaning procedures for non-dedicated equipment. Lack of demonstration of effectiveness of cleaning for dedicated equipment
	3. Pattern of failure to document investigation of discrepancies
	4. Pattern of failure to establish/follow a control system for implementing changes in the equipment
	5. Pattern of failure to qualify equipment, including computers
Materials System	1. Release of materials for use or distribution that do not conform to established

specifications

2. Pattern of failure to conduct one specific identity test for components

3. Pattern of failure to document investigation of discrepancies

4. Pattern of failure to establish/follow a control system for implementing changes in the materials handling operations

5. Lack of validation of water systems as required depending upon the intended use of the water

6. Lack of validation of computerized processes

Production System	1. Pattern of failure to establish/follow a control system for implementing changes in the production system operations
	2. Pattern of failure to document investigation of discrepancies
	3. Lack of process validation
	4. Lack of validation of computerized processes
	5. Pattern of incomplete or missing batch production records
	6. Pattern of non-conformance to established in-process controls, tests, and/or specifications
Packaging and Labeling System	1. Pattern of failure to establish/follow
	2. Pattern of failure to document investigation of discrepancies
	3. Lack of validation of computerized processes
	4. Lack of control of packaging and labelling operations that may introduce potential for mislabelling
	5. Lack of packaging validation
Laboratory Control System	1. Pattern of failure to establish/follow a control system for implementing changes in the laboratory operations
	2. Pattern of failure to document investigation of discrepancies
	3. Lack of validation of computerized and/or automated processes
	4. Pattern of inadequate sampling policies
	5. Lack of validated analytical methods
	6. Pattern of failure to follow approved analytical procedures
	7. Pattern of failure to follow an adequate OOS procedure

8. Pattern of failure to retain raw data

9. Lack of stability indicated methods

10. Pattern of failure to follow stability programs

6. STRATEGIC CMC TIPS FOR GMP COMPLIANCE

As indicated by Figure 7, apply the minimum CMC continuum to cGMPS. This graded nature is sometimes referred to as the 'spirit' of cGMPS during the early clinical development stages.

Figure 7. Graded nature of cGMPs during clinical development

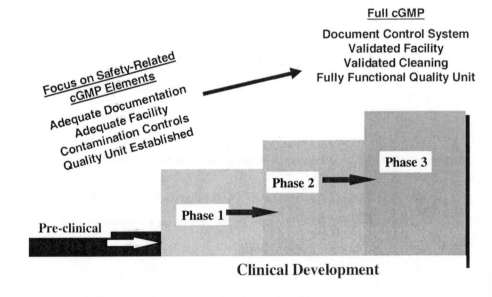

Take cGMPs seriously, whether for market approved products or for Phase 1 clinical trial materials.

When initiating manufacturing of product for human clinical trials, ensure that a Quality Unit is in place and functioning. It is most important that that a company have quality assurance and quality control activities from the very earliest stages of development. This is the only way to ensure that cGMPs are being integrated in every aspect of the manufacturing process and product testing. Not only is this a requirement but it also makes good business sense.

It should be emphasized that this Quality Unit can be a significant issue for the typical biopharmaceutical company. Most biopharmaceutical companies are 'start-ups' and 'one-product' companies. For these companies, the challenge will be to find adequate QA and QC staff both to develop the quality systems (which probably do not exist since the company is a R&D company) and to carry out the necessary functions. From my experience, these companies typically 'convert' (usually overnight) their process development and analytical staff members into 'quality staff members'. Here is where the use of an outside CMC consultant can be invaluable to provide the necessary depth of understanding to help the company realize its goal in a cost-effective manner.

For Phase 1 clinical trial materials, do all that is necessary to protect the biopharmaceutical product. As will be shown in subsequent chapters, this protection takes on different forms depending upon the specific needs of the manufacturing process. For example, the controls and monitoring needed to achieve an adequately controlled environment for a cell culture process using stainless steel bioreactors in a clean room is vastly different than that needed for bioengineered animals or plants growing in outside farms or fields.

Never take short cuts when it comes to safety of the product. Ensure that adequate contamination and cross-contamination controls are in place and effective.

As clinical development proceeds for the biopharmaceutical, fill-in the development of the quality systems and tighten process controls, that were established at the Phase 1 stage. Start thinking about what it will take to have all of the quality systems matured to meet the market approved expectations.

Prior to submitting the market approval dossier, ensure that all cGMP requirements are in being met for the facility, the utilities, the process, the product, and the personnel. Remember, one of the questions that will be asked by the FDA at the pre-BLA/NDA meeting is 'Are all of you facilities ready for an inspection at the time of the BLA/NDA submission?'[17]

Protect the product! Protect the patient! Protect your corporate reputation with the FDA! Protect your public reputation with your shareholders!

One final thought. While it takes only 297 words for the Ten Commandments and only 463 words for the Bill of Rights, why does it take over 10,000 words to describe the cGMPs requirements? Is it any wonder that it can get a bit confusing.

5

Recombinant Source Material: Master/Working Banks

"If there are any reasons to believe the manufacturing or controls for the clinical trial product present unreasonable health risks to the subjects ... such as a poorly characterized master or working cell bank."

CMC reason for the FDA to place a 'clinical hold'
on a submitted Phase 1 IND[37]

"Basis for a refusal to file under 21 CFR 601.2(a) ... insufficient description of source material (including characterization of relevant cell banking systems) ...."

CMC reason for the FDA to 'refusal to file'
a submitted BLA/NDA[31]

1. NEEDED: RELIABLE, CONTINUOUS, STABLE GENETIC SOURCE

The genetic source material needed to initiate a biopharmaceutical manufacturing process must be reliable, continuous and stable.

The genetic source material must be reliable in terms of safety and quality so that the desired specific biopharmaceutical product, and only that product, is manufactured, without the introduction of adventitious agents that could harm the patient.

The genetic source material must be continuous in terms of an adequate long-term inventory in order to be able to produce the same biopharmaceutical product over and over again, for extended periods of time.

The genetic source material must be stable in terms of storage so that the biopharmaceutical product produced during the first manufacturing campaign is the same product, both in terms of quality and quantity, as the product produced during the last manufacturing campaign.

When working with a recombinant living host, it cannot be emphasized enough that a manufacturer does not have a truly useable process until a reliable, continuous, stable source of starting material is in hand. Unfortunately, some companies learn the hard way and try to move forward either with a host of very low productivity or with an unstable host. They may be able to manufacture enough product to initiate the clinical development process, but soon enough they will find out that either the manufacturing costs are unacceptable for a commercial process or an inconsistent product quality keeps them away from receiving regulatory agency approval.

1.1. Three Primary CMC Concerns for Banks

To provide this reliable, continuous stable genetic source, a manufacturer will prepare a bank. The 'master bank' is a single pool of recombinant organisms prepared from a selected clone, aliquotted and stored under defined conditions. The master bank aliquots provide the characterized common starting source material for production. By the expansion of a master bank aliquot, a 'working bank' is prepared. The working bank can then be used for initiating production of the biopharmaceutical.

Regulatory agencies have great concern about the quality and safety of the master and working banks used to produce biopharmaceuticals. Their three primary CMC concerns can be summarized as follows:

1. The quality and safety of the banks impacts the quality and safety of the product

   The old adage of 'garbage in, garbage out' applies here.

2. The quality and safety of the banks cannot be fully assessed without adequate, extensive documentation

   The origin and handling of all of the genetic components used to assemble the recombinant organism directly impacts the master bank subsequently prepared from it.

3. The quality and safety of the banks cannot be fully assured without appropriate controls on all aspects of the bank preparation and extensive characterization of the bank

Quality and safety cannot be tested in, but must be designed in, especially procedures to prevent the entry adventitious agents.

It is the responsibility of each company to demonstrate to the regulatory agencies that their specific master/working bank is safe and appropriate for use to manufacture the desired biopharmaceutical product. This is definitely one place where 'trust me' does not work with the FDA.

1.2. Genetic Construction of a Bank

To genetically modify a living system to produce a biopharmaceutical, recombinant DNA technology is applied. It involves the systematic arrangement and manipulation of specific segments of nucleic acid for construction of composite molecules which, when placed into an appropriate host environment, will yield the desired product. The basic schematic of the genetic construction involved is illustrated in Figure 8.

Figure 8. Genetic construction of a recombinant organism

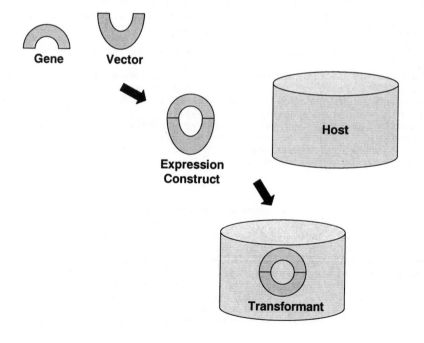

The 'gene' is the piece of DNA that codes for the sequence of the amino acids of the desired biopharmaceutical product. Many companies modify the codons of these genes in an attempt to improve the performance of some specific clinical effect of the produced biopharmaceutical (e.g., the humanization of the monoclonal antibody genes).

The 'vector' is a larger DNA molecule (frequently a plasmid) into which the gene of interest is inserted to yield the 'expression construct' (gene plus vector). The vector contains promoters (for efficient transcription), enhancers and other proprietary pieces of DNA that contain a company's tricks and trade secrets for allowing the biopharmaceutical product to be overproduced in the host.

The 'expression construct' is the vector into which the gene of interest has been inserted.

The 'host' is the parent living organism prior to insertion of the expression construct.

Once the expression construct has been inserted into the host it yields the 'transformant' (gene plus vector plus host).

To prepare the master bank, the transformant is cloned to ensure that a single genetic event is selected and passed on. The 'cloned transformant' is commonly referred to as the cell substrate. From the expansion of this cloned transformant, a master bank is prepared. Further expansion of a master bank aliquot leads to the working bank.

How extensive this genetic construction to prepare the banks is illustrated in Table 20, where the preparation of a master and working cell bank for a humanized monoclonal antibody is presented.

Table 20. Example of the genetic construction in preparing a master and working cell bank, using Palivizumab, a humanized monoclonal antibody[63]

Production and Selection of Murine Monoclonal Antibody 1129

BALB/c mice were immunized with human RSV strain A2 and boosted with vaccinia virus recombinant expressing RSV F protein or RSV strain A2. Spleens from these mice were fused with NS-1 murine myeloma cells and plated for individual colonies. Colonies which were positive for antibodies against RSV F protein underwent 2 rounds of limiting dilution to obtain 19 hybridoma lines. Ascites was prepared from these hybridoma lines and tested for F-protein binding sites using RSV mutants and competition studies. Three non-overlapping F-protein sites A, B, and C were identified using these antibodies. Antibodies against sites A and C appeared conserved among 23 clinical isolates. Eight of these antibodies were evaluated by MedImmune by intranasal prophylaxis followed by RSV challenge at 24 hours. MAb-1129 was one of the more effective antibodies

⇩

Humanization of MAb 1129

Messenger RNA from the hybridoma line MAb-1129 was purified and cDNA was generated. The V_H and V_L genes were amplified from the cDNA by PCR and cloned into the pUC18 vector. The cloned variable regions were sequenced and the predicted amino-acid sequences were compared to known human variable region sequences. The human V_L with the greatest homology was K102 with Jκ-4 and the human V_H with the greatest homology was with Cor and Cess (closely related V_H-II family members). Models, substituting these sequences into the coordinates of the MCPC603 crystal structure, were used for designing the CDR grafted human variable regions. The CDR grafted V_H gene was generated by assembly of synthetic oligonucleotides using PCR. The humanized V_L gene was generated by site directed mutagenesis of another CDR grafted V_L region.

⇩

Construction and Selection of the Expression NS0 Cell Line

NS0 cells are glutamine auxotrophs, with an absolute requirement for added glutamine or a glutamine synthetase gene. The use of glutamine deficient media allows for selection of cells transfected with expression

vectors containing the glutamine synthetase gene. Then the use of methionine sulphoxiamine (MSX), a glutamine synthetase inhibitor allows for the selection of NS0 cells with high copy numbers of the expression vector. The increased copy number also leads to increased production of product by the transfected NS0 cells. NS0 cells were grown and passaged. Palivizumab-expressing plasmid was linearized and used for the electroporation.

⇩

Selection of Candidate Cell Line and Adaption to Serum-Free Medium

The transfected NS0 cells were selected in glutamine deficient medium supplemented with 10% FCS. 147 colonies were analyzed by ELISA for human antibody. The 7 highest producing colonies (in pg/cell/day) were selected in MSX.

⇩

Single-Cell Cloning

The best of 21 subclones was selected and adapted to serum free medium and cloned by limiting dilution twice in the serum free medium. After the second round of cloning, the best cell line was selected. This clone was expanded to generate five vials of an Accession Cell Bank (ACB)

⇩

Master Cell Bank (MCB)

A vial of the ACB was expanded through a series of T-flasks and spinner flasks to generate the MCB. Cells from the spinner flask cultures were centrifuged and resuspended in a serum-free freezing medium to establish the MCB.

⇩

Working Cell Bank (WCB)

The MCB was further expanded to generate a WCB. Cells in one MCB vial were expanded through a series of T-flasks and subsequently, spinner flasks. Cells from the spinner flask cultures were centrifuged and resuspended in a serum-free freezing medium to establish the WCB.

2. SO MANY HOSTS TO CHOOSE FROM

To accomplish all of this, a manufacturer of a biopharmaceutical product will prepare (either in-house or outsource) a master and working bank(s) of the genetically modified living organism to be used in its production. But which host is to be genetically modified? The selection criteria will be based on a combination of scientific soundness and business needs.

2.1. Bank Terminology

Depending upon the type of host, the resulting master and working bank may consist of cells, viruses, transgenic animals or transgenic plants. The following terminology refers to the specific banks:

Recombinant Source	Master Bank	Working Bank
Cells (e.g., bacteria, yeast, mammalian, human)	Master Cell Bank MCB	Working Cell Bank WCB
Viruses (e.g., gene therapy retroviruses and adenoviruses; virus mediated transient plant transgenics)	Master Virus Bank MVB	Working Virus Bank WVB
Transgenic Animal	Master Transgenic Bank MTB	Working Transgenic Bank WTB
Transgenic Plant	Master Seed Bank MSB	Working Seed Bank WSB

Whatever the banking system is called, the end goal is to ensure that the desired biopharmaceutical product can be consistently manufactured for an extended period of time.

2.2. Choosing the Host

What is the 'ideal' host for the master/working banks?   One would think that after twenty plus years of biopharmaceutical manufacturing that this question could be easily answered.   But today, companies can choose between various types of cells (bacteria, yeast, plant, insect, mammalian, human), and various types of whole animals and plants as production systems for a biopharmaceutical.

2.2.1. Drivers to Reach a Decision

So then, why is one host chosen by a biopharmaceutical company over all of the others?   At least 4 drivers can be in play as a company reaches its decision:

1. Type of biopharmaceutical product needed

    Need for post-translational modifications or complex tertiary refolding and conformation shaping

    Need for higher recovery yields or opportunity for rapidly increasing capacity

2. Technology or corporate proprietary knowledge resident within the company

    Experiences of existing technical staff

    Patent ownership

3. Prior capital investment

    Existing investment in facilities and utilities

    Existing process equipment (e.g., bioreactor for mammalian cells versus a fermentor for bacterial cells)

4. Corporate image

    Technology platform communicated when raising venture capital funds

As can be seen, some of the drivers are science-based, others are financial-based. Fortunately for most biopharmaceutical products, a wide variety of hosts can be effectively used for their production.

2.2.2. Why Choose Recombinant Cells?

The choice of the host for the genetic construction does impact the CMC issues that have to be addressed for the master/working banks and subsequent manufacturing. The chosen organism will have certain inherent CMC advantages and disadvantages. The following are some advantages and disadvantages with the 4 most common cell types used in master and working cell banks:

- Recombinant Bacterial Cells

    o   Prokaryotes; simple organisms needing simple nutrients

    o   Cell walls resist damage, even when packed closely together

    o   Predictable liquid, stirred manufacturing processes

    o   High yields of expressed proteins

    o   Limit on size of DNA that can be cloned into host

    o   No post-translational modifications of expressed proteins (e.g., glycosylation)

    o   Most expressed proteins will need to be refolded to correct conformation

    o   Extensive regulatory agency experience (especially with *E. coli*)

- Recombinant Yeast Cells

    o   Eukaryotes; mitochondria perform much of the metabolic work

    o   Cell walls resist damage, even when packed closely together

    o   Predictable liquid, stirred manufacturing processes

    o   Moderate yields of expressed proteins

    o   Limit on size of DNA that can be cloned into host

    o   Some post-translational modifications of expressed proteins

    o   Presence of hyper-mannosylated forms of the expressed proteins

    o   Regulatory agency experience (especially with *S. cerevisiae*)

- Recombinant Insect Cells

  o Utilizes recombinant viruses and insects or insect cells

  o Recombinant baculovirus replicates only in cells of lepidopteran insects (e.g., butterflies and moths)

  o High yields of expressed proteins

  o Many post-translational modifications of expressed proteins

  o Requires stable stock of recombinant insect virus

  o Limited regulatory agency experience (no market approved therapeutic products yet)

- Recombinant Mammalian and Human Cells

  o Most common biopharmaceutical host

  o Cells replicate slowly, require complex nutrients and do not grow as well at high density because of waste accumulation and oxygen stress

  o Moderate yields of expressed proteins

  o No limit on size of DNA that can be cloned into host

  o Post-translational modifications of expressed proteins

  o Correct folding of expressed proteins

  o Proteins secreted into culture medium

  o Extensive regulatory agency experience (especially with hamster and mouse cells)

2.2.3. Why Bioengineered Animals or Plants?

The chosen animal or plant will have certain inherent CMC advantages and disadvantages. The following are some advantages and disadvantages with the use of transgenic master and working banks:

- Potential to produce enormous amounts of protein

- Freedom from stainless steel; ability to rapidly adjust capacity

- Environmental containment concerns

- Living animals and plants susceptible to pathogens

- Novel glycoforms from plants

- Very limited regulatory agency experience (no market approved therapeutic products yet)

3. CMC GUIDANCE ON PREPARATION OF A BANK

Regardless of the type of host selected for a specific biopharmaceutical, regulatory agencies require all manufacturers to accurately and adequately describe the preparation of their master/working banks, and they also require the manufacturers to appropriately and sufficiently characterize them.

According to Title 21 Code of Federal Regulations Part 610.18(c), 'Cell Lines Used for Manufacturing Biologic Products' shall be:

"(1) Identified by history,
(2) Described with respect to cytogenetic characteristics and tumorigenicity,
(3) Characterized with respect to in vitro growth characteristics and life potential, and
(4) Tested for the presence of dectable microbial agents.

Tests that are necessary to assure the safety, purity, and potency of a product may be required by the Director, Center for Biologics Evaluation and Research."[64]

The official regulations are very general and very limited considering all of the different types of recombinant organisms that can be used to produce biopharmaceuticals. For this reason, the regulatory agencies have had to rely on guidances to the industry to address the CMC needs for the diverse array of hosts and their resulting manufacturing processes.

3.1. Accurate and Thorough Description of Preparation

It is recognized that no bank testing regimen is able to detect all potential contaminants. Quality cannot be tested in. Documentation of proper control procedures used during assembling of the genetic components and preparation of the banks is important to provide reasonable assurance of true identity of these components and the absence of introduced contamination. Therefore, it is not surprising that the regulatory agencies require manufacturers to provide an accurate and adequate description of this history and preparation for their own assessment of safety of the prepared bank.

Because of the diversity of banks in use, the regulatory agencies have had to provide specific, detailed guidance on the CMC content necessary for describing the history and preparation of each type of bank used to produce biopharmaceuticals.

3.1.1. Recombinant Cell Banks

Table 21 presents some of the CMC documentation needed for many types of recombinant cell banks including those derived from bacterial, yeast, fungal, mammalian, avian, insect and human cells.

Table 21. CMC documentation required for describing the history and preparation of master/working cell banks used for production of biopharmaceuticals[40, 65]

Host	A description of the source, relevant phenotype, and genotype should be provided for the host cell used to construct the biological production system.
	The source of host cells (laboratory or culture collection) should be stated, and relevant references from the scientific literature should be cited. Information obtained directly from the source laboratory is preferred. When this is not available, literature references may be utilized.
	Characteristics of the original cells are required: • For human cell lines describe: tissue or organ of origin, ethnic and geographical origin, age, sex and general physiological condition; if known, the state of health or medical history of the donor should be reported along with the results of any tests of the donor for pathogenic agents • For animal cell lines describe: species, strains, breeding conditions, tissue or organ of origin, geographical origin, age and sex, the results of tests for pathogenic agents, and general physiological condition of the original donor • For microbial cells describe: species, strain, and known genotypic and phenotypic characteristics; pathogenicity, toxin production and other biohazard information, if any
	The cultivation history of the cells should be documented. The method originally used for the isolation of the cells should be described as well as the procedures used in the culturing of the cells in vitro and any procedures used to establish cell lines. A description of any genetic manipulation or selection should be provided.
	All available information regarding the identification, characteristics and results of testing of these cells for endogenous and adventitious agents should be provided.
	The results of the characterization of the host cell for phenotypic and genotypic markers, including those that will be monitored for cell stability, purity, and selection should be included.

Gene	A detailed description of the gene which was introduced into the host cells, including both the cell type and origin of the source material, should be provided.
	A description of the method(s) used to prepare the gene construct and a restriction enzyme digestion map of the construct should be included.
	The complete nucleotide sequence of the coding region and regulatory elements of the expression construct, with translated amino acid sequence, should be provided, including annotation designating all important sequence features.
Vector	Detailed information regarding the vector and genetic elements should be provided, including a description of the source and function of the component parts of the vector, e.g., origins of replication, antibiotic resistance genes, promoters, enhancers.
	A restriction enzyme digestion map indicating at least those sites used in construction of the vector should be provided.
	The genetic markers critical for the characterization of the production cells should be indicated.
Expression Construct	A detailed description should be provided of the cloning process which resulted in the final recombinant gene construct. The information should include a step-by-step description of the assembly of the gene fragments and vector or other genetic elements to form the final gene construct.
	A restriction enzyme digestion map indicating at least those sites used in construction of the final product construct should be provided.
Transformants/ Cloned Transformant	For recombinant products, the cell substrate is the cloned transformant. During the generation of the cell substrate, one or more of the following methods may be used: cell fusion, transfection, selection, colony isolation, cloning, gene amplification, and adaptation to specific culture conditions or media. Information regarding the methodologies utilized in developing the cell substrate helps to provide a clear understanding of the history of the cell substrate.
	The methods utilized to transfer a final gene construct or isolated gene fragments into its host, the mechanism of transfer, copy number, and the physical state of the final construct inside the host cell (i.e., integrated or extrachromosomal), should be provided.
	In addition, the amplification of the gene construct, if applicable, selection of the recombinant cell clone, and establishment of the seed should be completely described.
Master/ Working Banks	The preparation of the MCB and the WCB should be described in detail, including methods, reagents and media used, date of creation, quantity of the cell bank, in-process controls, and storage conditions.
	A description should be provided of the type of banking system used, the size of the cell bank(s), the container (vials, ampoules, or other appropriate vessels) and closure system used, the methods used for preparation of the cell bank(s) including the cryoprotectants and media used, and the conditions employed for cryopreservation and storage.

A description should be provided of the procedures used to avoid microbial contamination and cross-contamination by other cell types present in the laboratory, and the procedures that allow the cell bank containers to be traced. This should include a description of the documentation system as well as that of a labelling system which can withstand the process of preservation, storage, and recovery from storage without loss of labelling information on the container.

Manufacturers should describe their plans for protection from catastrophic events (e.g., fires, power outages and human error) that could render the cell bank unusable.

Results of the characterization of the MCB and WCB should be included.

The starting point of reference for estimates of in vitro cell age during manufacturing should be the thawing of one or more containers of the MCB.

3.1.2. Transgenic Banks

Table 22 presents some of the CMC documentation needed for for biopharmaceuticals produced from banks involving bioengineered animals and bioengineered plants. Note, that there is a similar amount of documentation required as that for the recombinant cell banks.

Table 22. CMC documentation required for describing the history and preparation of master/working transgenic banks[66, 67]

	Animal Based Biopharmaceuticals	Plant Based Biopharmaceuticals
Host	The history of the animals which donate gametes or embryonic stem cells (donor) and foster or recipient animals should be presented in detail and should include the: • species • breed • country of origin • general health • other available genealogical information Animals should have detailed veterinary evaluations of health, including specific tests for species and breed-related disease problems.	Thorough description of the host plant biology that includes information necessary to identify it in the narrowest taxonomic grouping applicable (e.g., genus, species, subspecies, variety or cultivar, line of designation) Description of the reproductive biology of the unmodified plant and production practices with regard to: • growth habitat as an annual, perennial, or biennial; • timing of sexual maturity and duration of flowering; • seed production and harvesting; • recognized practices for maintaining seed stock purity;

The use of animals from countries where transmissible spongiform encephalopathy (TSE) agents have been identified to have affected the species should be avoided. For control of adventitious agents, the donor and foster animals should meet the same criteria as required for adventitious agent screening required for admission to the production herd.

- conditions of growth;
- timing of harvest;
- method of harvesting;
- transporting, storage and sorting of harvested materials

Description of the host plant including levels of any toxins, anti-nutrients, and allergens known to be produced by the plant species and whether it is known to accumulate heavy metals

Statement on whether the plant is of a species used for food or feed in a raw or processed form

Gene

Detailed description of the original gene intended for introduction into the animals should be provided.

The natural protein and its functions should be described along with a description of its pattern of expression.

The methods used to clone and isolate the gene should be reported.

Description of the transgene should include properly scaled maps and all nucleotide sequences either previously published or newly determined.

Description of large segments of DNA in vectors should include detailed restriction maps if the entire nucleotide sequence has not yet been determined. In these cases, nucleotide sequences of cDNAs should be determined.

A detailed description of the gene which was introduced into the host cells, including both the cell type and origin of the source material, should be provided.

A description of the method(s) used to prepare the gene construct and a restriction enzyme digestion map of the construct should be included.

The complete nucleotide sequence of the coding region and regulatory elements of the expression construct, with translated amino acid sequence, should be provided, including annotation designating all important sequence features.

Vector

The vector sequence should be extensively characterized by restriction maps and nucleotide sequences.

The origin and characteristics of regulatory elements are of particular interest and should be carefully described.

Information of transcriptional control of the transgene including elements such as enhancers, promoters, suppressors, and presence or absence of dominant control regions should be reported if these elements are included in the transgene construct and will contribute to targeted expression.

Detailed information regarding the vector and genetic elements should be provided, including a description of the source and function of the component parts of the vector, e.g., origins of replication, antibiotic resistance genes, promoters, enhancers.

A restriction enzyme digestion map indicating at least those sites used in construction of the vector should be provided.

In addition, introduction of novel transcription factors should be fully described.

Expression Construct

The strategy used to create the final transgene construct should be described in detail.

The transgene construct should be extensively characterized by restriction maps and nucleotide sequences.

Full characterization of the recombinant DNA constructs or viral vectors used to transfer genes, including:

- Origin and function of all component parts of the construct, including coding regions (full-length and truncated sense constructs, antisense constructs, and constructs containing ribozymes, regardless of whether or not the coding region is designed or expected to be expressed in the bioengineered plant), antibiotic- or herbicide-resistance genes, origins of replication, promoters and enhancers
- Physical map of the construct(s) illustrating the position of each functional component
- Method used for plasmid propagation
- Sequences required for bacterial expression of plasmid constructs
- Nucleotide sequence of the intended insert up to and including the junctions at the 5'- and 3'-ends
- Any changes in codons to reflect more acceptable codon usage in plants

Transformants/ Cloned Transformant

The methods used to introduce recombinant DNA into animals should be described in detail. For example, all procedures used during generation of animals with germline alterations should be presented including techniques used in:

- isolation of the ova
- in vitro fertilization
- microinjection of blastula or of embryonic stem cell line
- embryo development and transfer
- other established or novel techniques

The methods used to generate animals with somatic alterations should also be described in detail.

The methods to be used for identifying founder transgenic animals and transgenic

Describe the gene transfer method in detail, and provide copy of relevant references.

Analysis of gene copy number, number of integration sites and demonstration of complete or partial copies inserted into the plant's genome.

Nucleotide sequence of the insert from DNA or mRNA retrieved from the stably-transfected plants to confirm integrity and fidelity of the DNA insert.

If a fragment of a coding region designed to be expressed in a plant is detected, determine whether a fusion protein could be produced and in which host tissues it may be located

If the transformation system utilizes a pathogenic organism or nucleic acid sequences from a pathogen, provide a description of the

animals in subsequent generations should be reported. The yields of the desired product within accepted criteria should be presented in detail. The yields of the desired product should be reported as well as any seasonal, age-related, or other variations in production. There should be verification of transgene expression at the expected tissue sites and/or at appropriate times in the host's life. This tissues should be carefully monitored and verification of normal processing should be undertaken.

pathogen, the strain, and the gene(s) involved; also include a description of any helper plasmids or analogous DNA fragments.

Complete description of the transformation process, including selection methods for the final transformant.

Describe in detail any tests used to evaluate the transformations process and provide the results.

Master/ Working Banks

Since animals cannot be stored indefinitely like cells, it is important to develop approaches to ensure that a desired product form a useful transgenic founder animal remains available for an extended period of time. This approach should take into account the possibility that genes from the transgenic animal may interact with different genetic backgrounds in the breeding partner, potentially affecting the quantity, quality, and purity of the product produced in each offspring.

If animals are used, the Master Transgenic Bank and the Working Transgenic Bank could consist of a limited number of highly characterized transgenic animals derived from a particular founder.

Description includes:
- Identification
- Method of production
- Results of analytical tests used to characterize it
- Size of the bank
- Storage conditions
- Data demonstrating viability, bioburden (including speciation of contaminants), uniformity of gene content, and stability

For all inserted coding regions, demonstration that the protein is produced or not produced (describe assay method and indicate limit of detection) as intended in the expected tissues consistent with the associated regulatory sequences driving its expression

Pattern and stability of inheritance and expression of the new traits over several generations.

3.2. Why Does The FDA Want So Much CMC Documentation?

As one can see, the list of CMC information required for describing the history and preparation of each type of bank is quite extensive. The regulatory agencies require knowing where the genetic elements (the gene, the vector, the host) came from, how they were modified, how they were linked together to produce the transformant, and how all of this was controlled. They also require that these genetic elements be extensively characterized (in addition to the characterization required on the final master bank). Furthermore, a detailed description of the procedures and methodologies used to manufacture the bank is required.

I am often asked why do the regulatory agencies want to know all of this information? The primary reason is for them to be able to properly assess the safety implications for the prepared banks. As stated in the ICH guidance document on 'Quality of Biotechnological/Biological Products – Derivation and Characterization of Cell Substrates (ICH Q5D)'[65]:

> "Events during the research and development phases may contribute significantly to assessment of the risks."

> "The information supplied is meant to facilitate an overall evaluation which will ensure the quality and safety of the product."

It is the responsibility of the company to provide this required information, accurately and adequately, to them.

3.3. When is Full CMC Documentation Needed?

How much documentation is necessary? At the Phase 1 IND stage, the regulatory requirement is for a 'brief' CMC description, while at the market approval stage (the filing of the BLA/NDA) the requirement is for a 'full' CMC description of the history and preparation of these banks. For bank characterization, the FDA typically prefers to see copies of the characterization reports included in the IND. The characterization reports are required for the BLA/NDA.

Although the full CMC descriptions are not required until the BLA/NDA dossier is submitted, I usually recommend to companies that they prepare this CMC information and submit it anyway in their IND to initiate the Phase 1 clinical trials. The inclusion of this information serves four purposes:

(1) Forces a company to record what occurred

> The master bank, and possible the working bank, are already manufactured. The master bank should rarely, if ever, be prepared again. What better time to secure written documentation of what actually happened than close to the time it occurred. It won't be any easier to compile this documentation in the future, in fact, with people coming and going in companies, some CMC information may actually get lost if not compiled early.

(2) Forces a company to review what occurred

> When sometime is being written up, it also is reviewed. This can be a valuable time to ensure that nothing has been missed, or to identify any concerns with the traceability of the origins of the host, vector and gene of interest, or to determine if there is an anomaly with any of the test results. Thus, time is available to address or correct any issues. Preparing the market approval dossier is not the time to find a surprise.

(3) Serves as an excellent repository of this information

> When the BLA/NDA dossier is prepared in the future, the CMC content needed will already be present in the IND submissions, where a regulatory affairs group can locate it.

(4) Can detect a problem earlier and allow for an easier investigation

> Unfortunately, a number of biopharmaceutical companies have experienced the sad story of starting production with what they believed to be the bank for their specific biopharmaceutical only to find out that somehow things got mixed up during preparation and the product was not exactly what they intended to produce. Accurate and adequate documentation helps provide assurance that all the pieces used in the genetic engineering process are the correct ones.

3.4. What If CMC Documentation is Missing?

How important is it that all of this description for the master/working banks be documented? The regulatory agency preference is that careful records of the origin, identity and handling of each genetic component and the assembly of the expression construct, and the preparation of the resulting banks be thoroughly documented.

However, deficiencies in documented history may not, by itself, be an impediment to product approval. As indicated in ICH Q5D:

> "Careful records of the manipulation of the cell substrate should be maintained throughout its development. Description of cell history is only one tool of many used for cell substrate characterization. In general, deficiencies in documented history may not, by itself, be an impediment to product approval, but extensive deficiencies will result in increased reliance on other methods to characterize the cell substrate."[65]

Therefore, extensive deficiencies in documentation will result in increased reliance on the characterization methods, and this most likely will also generate some extensive discussion with the regulatory agency. Personally, this is not a comfortable position to be in either with a FDA reviewer. It is far better to do it right at the start.

3.5. Don't forget GMPs During Preparation of the Bank

While the preparation of the banks are not specifically required to be under GMP controls according to ICH Q7A, their preparation must be controlled to prevent either the entry of adventitious agents or cross-contamination. Don't forget GMPs, or more accurately 'appropriate process controls', during the manufacture and storage of the master/working banks[22]:

- Access to banks should be limited to authorized personnel

- Banks should be maintained under storage conditions designed to maintain viability and prevent contamination

   For recombinant cells, proper storage of the cells is the frozen state. The two main enemies of successful cell freezing are intracellular ice crystal formation and intracellular dehydration, both of which can impact the functional recovery of the cell.   Typical freezing procedures include growing the cells in a medium similar to the production medium, to a cell density of ~ 10^8/mL for bacterial and yeast cells, and ~ 10^6/mL for animal and human cells. Slow freezing (~1 °C per minute) in the presence of a cryoprotectant (e.g., DMSO, glycerol, fetal calf serum or other protein), with storage below the -130 °C glassy transition temperature of water (storage in the vapour phase of liquid nitrogen, -196 °C) are used to protect the cells.

   Control over the storage of the banks is a critical concern to the regulatory agencies. Brief exposures to temperatures above -130 °C (either in storage or shipment) may adversely affect the cells. Storage in the vapour phase of liquid nitrogen is preferred since liquid nitrogen can enter into spaces between the cap and body of plastic storage vials or into tiny cracks in glass ampoules.

   For transgenics, room temperature storage of the transgenic bank (animals) or the transgenic seeds (plants) limits their useful lifetimes. For this reason, these banks will need to be regenerated over time, with controls to ensure that the fidelity or yield of the expressed biopharmaceutical product is not impacted.

- Records of the use of the bank vials should be maintained

   Careful organization of the banks is essential as many facilities have multiple banks stored together. To prevent mishandling, any removed master/working bank vial should be identified for lot and vial number by two workers and recorded in the batch record.

- Banks should be periodically monitored to determine suitability of use

4. CMC GUIDANCE ON CHARACTERIZATION OF A BANK

It is important to prevent a contaminated bank from being used in production and to avoid a loss of product availability or development time resulting from the need to recreate a bank found to be unusable. The regulatory agencies require an appropriate and sufficient characterization of the prepared master/working bank.

4.1. Appropriate and Sufficient Characterization

The characterization and testing of banks is a critical component of the control for biopharmaceuticals. Characterization allows the manufacturer to assess the bank with regard to presence of other recombinant and non-recombinant hosts, adventitious agents, endogenous agents and molecular contaminants (e.g., toxins from the host organism). The objective of characterization is to confirm the identity, purity, and suitability of the bank for manufacturing use.

4.1.1. Six Key Elements for a Thorough Characterization

What is involved in an appropriate characterization of a bank? There are 6 key elements:

1. Identity : Presence of correct gene, vector and host

2. Purity: Absence of other hosts, absence of endogenous viruses, absence of adventitious agents (viruses, transmissible spongiform encephalopathy agents)

3. Suitability of Recombinant Organism: Host viability, host productivity of expressed protein, fidelity of expressed protein

4. Other Specific Safety Concerns: Specific issues for the recombinant organism chosen

5. Stability During Storage: Suitability of the recombinant organism over time during storage

6. Stability During Cultivation for Production: Suitability of the recombinant organism during the aging of the host

The specific testing program involving these 6 elements will vary for any given bank according to the biological properties of the specific recombinant organism (e.g., growth requirements), its history of preparation (including use of human-derived and animal-derived biological reagents) and available testing procedures.

The master bank is tested more rigorously than the working bank. Since the working banks are generated from characterized master banks, the extent of testing can be less, primarily in the amount of genetic identity testing and endogenous virus testing that needs to be performed (if endogenous viruses are not present in the master bank)..

Because of the diversity of banks in use, the regulatory agencies have provided specific, detailed guidance on the characterization necessary for some of the types of banks used to produce biopharmaceuticals.

4.1.2. Recombinant Cell Bank Characterization

Table 23 presents some of the characterization needed for many types of recombinant cell banks including those derived from microbial cells (bacterial, yeast, fungal) and metazoan cells (mammalian, human).

Table 23. Characterization required for master/working cell banks used for production of biopharmaceuticals[40, 65, 68]

	Microbial Cell Banks	Metazoan Cell Banks
Identity	For most microbial cells, analysis of growth on selective media is usually adequate to confirm host cell identity at the species level for the host cell bank and the transformed cell bank.	Morphological analysis may be a useful tool in conjunction with other tests. In most cases, isoenyme analysis is sufficient to confirm the species of origin for cell lines derived from human or animal sources; other tests may be appropriate depending on the history of the cell line.
	For *E. coli*, where a variety of strains may be used, biological characterization methods such as phage typing should be considered as supplementary tests of identity.	Other technologies may be substituted to confirm species of origin, including, for example, banding cytogenetics or use of species-specific antisera. An alternate strategy would be to demonstrate the presence of unique markers, for example, by using banding cytogenetics to detect a unique marker chromosome, or DNA analysis to detect a genomic polymorphism pattern (for example, restriction fragment length polymorphism, variable number of tandem repeats, or genomic dinucleotide repeats).
	Expression of the desired product is also considered adequate to confirm the identity of the microbial expression system.	
	Restriction endonuclease mapping or other suitable techniques should be used to analyse the expression construct for copy number, for insertions or deletions, and for the number of integration sites.	Either confirmation of species of origin or presence of known unique cell line markers is considered an adequate test of identity.
	For extrachromosomal expression systems, the percent of host cells retaining the expression construct should be determined.	Restriction endonuclease mapping or other suitable techniques should be used to analyse the expression construct for copy number, for insertions or deletions, and for the number of integration sites.
	Both the nucleotide sequence encoding the product and the amino acid sequence of the expressed protein should be verified.	Both the nucleotide sequence encoding the product and the amino acid sequence of the expressed protein should be verified.

Purity	The design and performance of specific tests for adventitious microbial agents and adventitious cellular contaminants in microbial cell banks should take into account the properties of the banked cell, the likely contaminants based upon scientific literature, source, methods and materials used for cultivation, and other organisms present in the banking laboratory.	Tests for the presence of bioburden (bacteria and fungi) should be performed on individual containers (1% of the total number but not less than two containers) of the MCB and WCB. In all other aspects, the current methodologies described in either the European Pharmacopoeia (Ph.Eur.), the Japanese Pharmacopoeia (JP) or the U.S. Pharmacopoeia (USP.) for testing microbial limits or microbial sterility may be considered adequate.
	Visual examination of the characteristics of well-isolated colonies is suggested, using several microbiological media, of which some do and do not support growth of the cell substrate.	Tests for the presence of mycoplasma should be performed on the MCB and WCB. Current procedures considered adequate include both the agar and broth media procedures as well as the indicator cell culture procedure. Current methods for mycoplasma testing are described in Ph. Eur., JP, and FDA Points to Consider in the Characterisation of Cell Lines Used to Produce Biologicals. Testing cells derived from a single container is generally considered adequate.
	However, it is not intended that manufacturers necessarily characterize resistant mutants of the cell substrate arising from such studies, or other artefacts of such assays. Rather, the purpose of such assays is to detect existing contaminants.	
	For bacterial cells, test for contamination with both lytic and lysogenic bacteriophages.	Virus testing of cell substrates should be designed to detect a wide spectrum of viruses by using appropriate screening tests and relevant specific tests, based on the cultivation history of the cell line, to detect possible contaminating viruses. Applicants should consult the ICH guideline on viral safety.
		The purity of cell substrates can be compromised through contamination by cell lines of the same or different species of origin. The choice of tests to be performed depends upon whether opportunities have existed for cross-contamination by other cell lines. Additional assurance of lack of cross-contamination can be provided by successful preparation of the intended biopharmaceutical product from the cell substrate.
Suitability	Viability and expression of the desired biopharmaceutical product demonstrate suitability of the recombinant organism.	
Other Specific Safety Concerns	None	Utilisation of karyology and tumorigencity testing for evaluating the safety or characterizing a new cell line may be useful depending on the cells, the nature of the product and the manufacturing process.
		For products that are highly purified and that contain no cells, karyology and tumorigenicity testing are generally not considered necessary,

provided that appropriate limits for residual host cell DNA are shown to be consistently met.

Repetition of tumorigenicity testing for cells with already documented evidence of tumorigenicity is not considered necessary.

Stability During Storage	Evidence for banked cell stability under defined storage conditions will usually be generated during production of clinical trial material from the banked cells.
	Data from the preparation of clinical materials will demonstrate that the revived cells can be used to prepare the desired product.
	In the case when production does not take place for a long period of time, viability testing on the cell bank should be performed at a described interval.
	If the viability of the cell substrate is not significantly decreased, generally no further testing of the MCB or WCB is considered necessary.
Stability During Cultivation	At least two time points should be examined, one using cells which have received a minimal number of subcultivations, and another using cells at or beyond the limit of in vitro cell age for production use.
	Consistency of the coding sequence of the expression construct should be verified in cells cultivated to the limit of in virtro cell age for production use or beyond by both nucleic acid testing and analysis of the purified protein product.

4.1.3. Example of Characterization of a Bacterial Cell Bank

An illustration of adequate and sufficient characterization for a bacterial cell bank of recombinant *E. coli* is provided in Table 24. The table illustrates the types of tests to be performed on the Master Cell Bank and on the Working Cell Bank. The table also illustrates the types of tests to be performed on the End of Production Cells (cells obtained at the end of the bioreactor production) or Cells Beyond Production (cells obtained several passages beyond the end of production; frequently carried out on small scale). This latter stage is designed to address the genetic stability concerns. Because bacterial cells do not harbour animal or human viruses, there is no need to test these banks for such adventitious agents; however, if biologically-derived raw materials are used there remains the possibility of introducing viruses into the bank.

Table 24. Illustration of cell bank characterization for an *E. coli* cell bank

Element	Test	MCB	WCB	EPC/CBP (Genetic Stability)
Identity	Gene: DNA Sequencing	X		X
	Vector: Plasmid Retention in Colonies	X	X	X
	Vector: Restriction Endonuclease Mapping	X	X	X
	Vector: Copy Number Determination	X	X	X
	Host: Morphology on Selected Media	X		
	Identity of Secreted Protein	X	X	
	Peptide Mapping of Purified Protein	X		X
Purity	Culture Purity: Colony Morphology and Consistency on Selected Media	X	X	X
	Absence of Bacteriophages	X	X	X
	Specific Bovine or Porcine or Human Viruses (only if animal- or human-derived components were used in manufacture of the bank)	X	X	
Suitability of Recombinant Organism	Cell Viability	X	X	X
Other safety concerns	N/A			

MCB = Master Cell Bank; WCB = Working Cell Bank; EPC = End of Production Cells;
CBP = Cells Beyond Production

4.1.4. Example of Characterization of a Mammalian Cell Bank

An illustration of adequate and sufficient characterization for a recombinant Chinese Hamster Ovary (CHO) cell bank is provided in Table 25. The table shows the additional testing required because these cells can harbour and amplify virus contaminants. Such extensive testing comes at considerable cost (depending upon the extent of viral testing, the cost can readily approach a quarter of a million dollars) and requires considerable time to complete the testing (sometimes up to 6 months by time the contract lab QA final report is available). Because of the complexity of the tests to be performed and the infrequency of need in any specific company, most bank characterization is outsourced to contract testing labs who run these assays all of the time (for example, BioReliance, www.bioreliance.com; Q-One Biotech, www.q-one.com; MDS Pharma Services, www.mdsps.com are just a few contract companies that offer this service). These contract testing companies can be invaluable sources for technical information on bank characterization and can guide a biopharmaceutical company in designing a proper characterization plan for their respective master/working bank.

Table 25. Illustration of cell bank characterization for a Chinese Hamster Ovary (CHO) cell bank

Element	Test	MCB	WCB	EPC/CBP (Genetic Stability)
Identity	Gene: DNA Sequencing	X		X
	Vector: Restriction Endonuclease Mapping	X	X	X
	Vector: Copy Number Determination	X	X	X
	Host: Isoenzyme Analysis	X		
	Identity of Secreted Protein	X	X	
	Peptide Mapping of Purified Protein	X	X	X
Purity	Sterility	X	X	X
	Mycoplasma – DNA Staining	X	X	X
	Mycoplasma – Direct Inoculation	X	X	X
	In Vitro Assay for Adventitious Viruses	X	X	X
	In Vivo Assay for Adventitious Viruses	X	X	X
	Hamster Antibody Production (HAP) Test	X		
	Extended S$^+$L$^-$ Assay for Viruses	X		
	Electron Microscopy for Retroviral Particles	X	X	X
	Reverse Transcriptase Assay for Retroviruses	X		X
	Specific Bovine or Porcine or Human Viruses (only if animal- or human-derived components were used in manufacture of the bank)	X	X	
Suitability of Recombinant Organism	Cell Viability	X	X	X
Other safety concerns	N/A			

MCB = Master Cell Bank; WCB = Working Cell Bank; EPC = End of Production Cells;
CBP = Cells Beyond Production

4.2. How Much Characterization and When?

From my experience, I recommend that companies complete as much characterization as possible of their master and working banks prior to the submission of the Phase 1 IND. The last thing a company wants to find out at a later date is a surprise with the quality of the bank, which might cause the company to pull the product from the clinic.

At a CMC minimum, three main questions about the banks will have to be addressed before a regulatory agency will allow a Phase 1 clinical trial to proceed:

1. Can I prove that the bank yields the intended biopharmaceutical product?

   At a minimum, there needs to be a match between the measured DNA sequence and the measured amino acid composition/sequence of the biopharmaceutical.

2. Can I prove that the bank is safe, specifically from viruses and TSEs?

   The FDA will want copies of all reports supporting the virus testing results included in the IND submission. Remember, safety is the number one criteria for the FDA to allow a clinical study either to proceed or not.

3. Can I prove that the bank is stable in order to manufacture the necessary clinical materials?

   Typically, the frozen storage stability of the master and working banks are determined during the manufacturing of the actual clinical trial materials. However, the FDA will have concerns if there are any early indications that the viability of the bank is decreasing significantly.

   To assure that the recombinant host was genetically stable at least through production, many companies perform end of production cell (EPC) testing on each bioreactor run.

For Phase 1, it is typically not possible either to know the shelf life of a bank nor to know whether the recombinant organism displays any genetic instabilities. However, prior to Phase 3, a company should have an understanding of both of these bank characteristics. If there is any concern about the viability of the bank over storage time, consider re-laying down the bank (i.e., expand an aliquot of the master bank to prepare new frozen aliquots). Genetic stability results determine the period over which a bank can be used. The limit of in vitro cell age used for production (which is a measure of the period between thawing of the master bank aliquot and harvest of the production measured by elapsed chronological time in culture, population doubling level or passage level) is based on data derived from production cells expanded under pilot-plant scale or commercial-scale conditions to the proposed in vitro cell age (end of production cells) or beyond (cells beyond production).

Insist on validated assays at the contract laboratory. Although the bank will be prepared prior to Phase 1 start, the bank will most likely be included in the BLA/NDA submission for market approval. In this submission, both the full reports on bank characterization and the validation data to support the methods used are included. This requirement for validated assays is clearly stated by the regulatory authorities:

"Analytical methods should be validated for the intended purpose of confirmation of sequence. The validation documentation should at a minimum include estimates of the limits of detection for variant sequences. This should be performed for either nucleic acid or protein sequence methods."[68]

"Virus titrations suffer the problems of variation common to al biological assay systems. Assessment of the accuracy of the virus titration and reduction factors derived from them and the validity of the assays should be performed to define the reliability of a study."[69]

Competent labs have this validation data and they can readily provide a validation summary. One doesn't want to have to repeat these expensive assays because of a lack of assay validation data.

4.3. Critical Concern for Virus Safety in Banks

Viruses are not wanted in the banks. Viruses could alter the normal growth and physiological behaviour of the recombinant living organism which in turn could impact the production of the desired recombinant protein leading to lower productivity and higher impurity levels. Viruses also present a significant potential safety concern to patients.

Viruses in banks can be endogenous. Endogenous viruses are part of the germ line of the host and covalently integrated into the genome. These viruses can be infective or defective (e.g., Chinese Hamster Ovary cells are well known to contain non-infectious defective retroviral particles). The infective viruses can be active or latent (i.e., virus that becomes active as the host ages). All recombinant living hosts have the possibility of containing an endogenous virus (e.g., bacteria can contain bacteriophages, plants can contain plant viruses, etc.).

Viruses in banks can also be adventitious. Adventitious viruses are unintentionally introduced viruses. The most common source of adventitious virus is the use of animal- or human-derived raw materials in the manufacture of the bank. Viruses derived from cows, pigs, goats, horses, humans and other living sources can enter into the bank through these raw materials. Bovine viruses can arise from the use of bovine serum in the medium, porcine viruses can arise from the use of porcine trypsin to dislodge cells, and human viruses such as HIV or hepatitis can arise from the use of human transferrin in the medium. It is thus most important, that if these biologically-based materials are needed in the manufacture of the bank, that they be carefully selected, tested and approved by the Quality Unit. An example of this rigorous control is illustrated in the EMEA guidance on the use of bovine serum in the manufacture of human biological products.[70]

The goal is to demonstrate that the bank is free of virus. This is the main reason that the regulatory agencies insist on such an extensive list of virus tests for banks. The type of virus testing required for these banks is host dependent. Bacterial and yeast host cells do not harbour viruses that are known to be pathogenic to humans. Insect host cells can harbour several of these viruses: Japanese encephalitis virus, dengue, Raft Valley fever, etc. But it is the human host cells that harbour the greatest number of potential pathogenic viruses: HIV I and II, Hepatitis B and C, Human Herpes viruses, Epstein Barr virus, etc.

Table 26 illustrates the extent of viral testing performed on a master and working cell bank of a market approved biopharmaceutical produced from a recombinant murine cell line, which apparently had contact with bovine-, caprine-, and human-derived materials during its preparation.

Table 26. Extensive viral testing performed for on both the master and working cell banks of the monoclonal antibody Remicade®, infliximab[71]

Mouse Antibody Production(MAP) Test (LCM, Ectromelia, GDVII, Hantaan, MVM, MAV, MHV, PVM, Polyoma, Reo-3, Sendai, EDIM, MCMV, K, Thymic, & LDVH)

Thymic Agent Virus Test

Bovine (Cow) Viruses (BVD, IBR, PI3, BPV & BAV-3)

Caprine (Goat) Viruses (CAV, CHV & CAEV)

Extended Assays (S⁺L⁻ Focus & XC Plaque)

Mus dunni Assay (ERV)

Extended Mus dunni Assay

Dunni Cell Co-Cultivation Assay

Extended MCF Test (Mink & SC-1)

Transmission Electron Microscopy (TEM)

Reverse Transcriptase (RT)

In vitro Adventitious Agents (HeLa, H9, MRC-5, Vero & C168J cells)

In vivo Adventitious Agents (suckling & adult mice, guinea pigs, embryonated hen eggs)

PBL Co-Cultivation Test

PCR (HIV 1 & 2, HTLV I)

DNA hybridization (HIV-1 & 2, HTLV I & II, HBLV, EBV, CMV, JCV, HBV, BIV, BLV, SIV, SRV)

But what happens if there is a virus present in the prepared bank? Can it still be used? The ICH Guideline on Viral Safety Evaluation of Biotechnology Products Derived from Cell Lines of Human or Animal Origin[69] provides the regulatory response. If the virus present is non-pathogenic and endogenous, the bank can be used, but only after extensive viral safety testing and evaluation. If the virus present is adventitious, the answer is typically 'No'. If the virus present is a human pathogen, the answer is a strong 'No'. Some of the viruses that could infect the typical mouse or hamster mammalian cell bank, and are known to be human pathogens, are presented in Table 27.

Table 27. Some viruses that could infect a mouse or hamster cell bank, known to be pathogenic to humans[69]

Mouse Cell Bank	Hamster Cell Bank
Lymphocytic Choriomeningitis Virus	Lymphocytic Choriomeningitis Virus
Sendai Virus	Sendai Virus
Hantaan virus	Reovirus Type 3
Lactic Dehydrogenase Virus	

As a final note, there are no known human viral pathogens harboured by plants; thus, for many biopharmaceutical products, transgenic plants represent a safer, less-expensive production host than mammalian cell substrates.[72]

4.4. Minimizing the Risk of TSEs

Transmissible spongiform encephalopathies (TSEs) include scrapie in sheep and goats, chronic wasting disease in mule deer and elk, bovine spongiform encephalopathy (BSE) in cattle, as well as Kuru and Creutzfeldt-Jakob Disease (CJD) in humans. Agents causing these diseases ('prions') replicate in infected individuals, generally without evidence of infection detectable by available diagnostic tests. After incubation periods of up to several years the agents cause disease and, finally, lead to death. No means of therapy are known. BSE was first recognized in the United Kingdom in 1986. It is clear that BSE is a food borne infection. There is convincing evidence that the new variant of CJD is caused by the agent that is responsible for BSE in cattle. The appearance of the new variant form of CJD has raised further concerns that the BSE agent can be transmitted to humans. Therefore, one can readily understand the caution that the regulatory agencies have placed on the use of animal- or human-derived raw materials in manufacturing of banks.

The FDA's concern about TSEs is not new. In 1991, the FDA issued to biological product manufacturers a letter requesting them to provide information regarding the source and control of any bovine- or ovine-derived materials used in preparing the products. They also were asked to provide a description of the testing performed on each lot of raw material.[73]   In 1996, the FDA issued another letter to biological product manufacturers instructing them not to use bovine-derived materials from cattle born, raised or slaughtered in BSE-countries.[74]   And then again, in 2000, the FDA issued another strong recommendation to the industry on this issue:

> "CBER strongly recommends that manufacturers take whatever steps are necessary to assure that materials derived from all specifies of ruminant animals born, raised or slaughtered in countries where BSE is known to exist, or countries where the USDA has been unable to assure FDA that BSE does not exist, are not used in the manufacure of FDA-regulatred products intended for administration to humans.   The Agency has previously recommended that manufacturers take the following steps to prevent this occurrence:
>
> 1.   Identify all ruminant-derived materials (e.g., culture medium, transferring, albumin, enzymes, lipids) used in the manufacture of regulated products.   FDA considers the manufacture of biological products to include the preparation of master (including the original cell line) and working cell banks, as well as materials used in fermentation, harvesting, purification and formulation of the products.
>
> 2.   Document the country of origin and all countries where the live animal source has resided for each ruminant-derived material used in the manufacture of the regulated product. The regulated-product manufacturer should obtain this information form the supplier of the ruminant-derived product. The certification of slaughter, as required by the country of origin of live animals, from the supplier.   Documentation should be maintained for any new or in-process lots of licensed, cleared or approved products; products pending clearance or approval; and investigational products intended to be administered to humans.
>
> 3.   Maintain traceable records for each lot of ruminant material and each lot of FDA-regulated product manufactured using these materials.   These records should be part of the product batch records and available for FDA inspection.   Such records should be maintained for products manufactured at foreign as well as domestic facilities."[75]

Could the FDA been any clearer in communicating their concern to the biological industry? Note in this letter, the FDA makes it very clear that how the parental cell line (i.e., the cloned transformant) was prepared is of concern to them.   Also note in this letter, the FDA emphasizes that the TSE concern applies just as much to clinical trial materials as to marketed products.

So it was a surprise in mid-2000, when CBER reported that several manufacturers of vaccines had not been following their requirements when preparing their master and working banks. The FDA uncovered that 6 marketed vaccines used bovine-derived materials from BSE-containing countries; they also uncovered that 2 marketed vaccines used bovine-derived materials from unknown sources, in part because manufacturers did not maintain or did not have access to records of the source of such materials. The net result of this disturbing discovery was that the FDA required each of these manufacturers to carry out the following:

> "Working bacterial and viral seed banks and working cell banks that were established using bovine-derived materials sourced from countries on the USDA list should be re-derived with bovine-derived materials from countries not on the USDA list.
>
> Master bacterial and viral seed banks established in a similar manner do not need to be re-derived; the potential risk presented by the master seed banks is even more remote that that presented by the working seed banks and is outweighed by the risk of altering the bacterial or viral vaccine through re-derivation."[76]

The companies either got themselves into this dilemma either because they lacked adequate CMC regulatory compliance or they did not keep current with updated CMC regulatory compliance (some of the master and working banks could have been made before the BSE concern was communicated by FDA). Fortunately, to date, there has been no reported BSE or vCJD infection due to any biopharmaceutical product. The regulatory agencies intend to keep it that way.

It should be noted that the EMEA in Europe has the same intense concern about the control of the use of the animal-derived and human-derived components, and they have issued their own guidance on minimizing the risk of TSEs.[77] European manufacturers also have had a problem with their cell banks for vaccines not meeting the BSE-restrictions.[78]

From a CMC regulatory compliance position, the manufacturer must take the responsibility today for justifying why they need to continue to use an animal-derived or human-derived raw material in their banks.

5. A SUCCESSFUL CMC STRATEGY FOR BANKS

If there was one place in a biopharmaceutical process where it is not wise to take shortcuts with the minimum CMC requirements, it would be at the preparation and characterization of the master and working bank. These banks are the source material for the recombinant organism and any problems here will continue through the rest of the manufacturing process. It is here that I would recommend, that to decrease manufacturing risk, that full documentation and full characterization be performed on the master bank prior to Phase 1.

For Phase 1 clinical studies, consider preparing only the master bank. This can cut down on the manufacturing costs and manufacturing time required to get a biopharmaceutical product into the clinic. However, unless there is a low utilization of the master bank aliquots, a working bank should be prepared prior to Phase 3 clinical studies. In this way, the biopharmaceutical product manufactured from the working bank can be incorporated into the pivotal clinical trials. Remember, the final goal is to have at the time of market approval at least a 10-20 year supply of the bank, with aliquots stored securely in multiple locations.

To ease the regulatory burden, consider selecting a host that the regulatory agencies have experience with. This will eliminate the unknown aspect of a new host system that the FDA will naturally be conservative toward in its review. Also avoid hosts with known endogenous infectious viruses. This will eliminate the extra hurdles that will have to be mastered in order to get clinical trial and market approval.

If at all possible, avoid the use of animal-derived and human-derived raw materials in the manufacture of the banks.

While the preparation of the banks are not specifically required to be under GMP controls according to ICH Q7A, their preparation still must be controlled to prevent both the entry of adventitious agents and cross-contamination. This can be a challenge for some companies at this early clinical development stage in that these banks may have been prepared in a non-controlled R&D laboratory before they have even established the Quality Unit to oversee the operation. In today's regulatory environment, this presents a significant risk to the company. If a company doesn't have a dedicated banking area under Quality Unit control, it would be prudent to contract out the manufacture of the bank to a competent laboratory.

Don't be lulled into thinking that a bank will only be used for early clinical studies. If clinical success comes quickly, one can be assured that the bank will be considered for the BLA submission. Therefore, apply the same standard to all banks.

The FDA will prevent a poorly characterized bank from being used to manufacture a biopharmaceutical for clinical trials. Count on it!

Production: Expansion of the Recombinant Organism and Expression of the Biopharmaceutical

'The harvested yield from the production bioreactor is
so low that the eventual commercial scale operation
would require in excess of the current
global bioreactor capacity'

A real life scenario frequently caused by lack of advanced planning
and/or by an insufficient investment in development
of the production process

1. GOALS: IDENTITY, CAPACITY AND CONSISTENCY

From an aliquot of the master/working bank, the number of recombinant organisms (i.e., cells, offspring, seeds, etc.) needs to be expanded in order to produce an adequate amount of the desired biopharmaceutical product. For recombinant living cells, the production process will require bioreactors to grow this adequate amount of cells. For transgenic animals or plants, the production process will require either breeding enough animal offspring or planting an adequate number of acres.

1.1. Two Major CMC Regulatory Concerns for Production

Regardless of the production process employed or its scale, the regulatory agencies have two major concerns because of the use of living organisms:

1. The production process must be adequately controlled and monitored to prevent adventitious agent contamination during the handling of the living recombinant organism

    > Living organisms are much more difficult to control than a chemical synthesis. Once infected with an adventitious agent, the possibility of amplification of that agent during the production process exists.

2. The production process must be adequately controlled and monitored to ensure that there is no unacceptable change to the identity, the quality or the safety of the expressed biopharmaceutical

    > Living organisms are susceptible to environmental changes during the production process which could impact the expressed product. Living organisms can undergo physiologically changes upon aging which also could impact the expressed biopharmaceutical.

Both of these regulatory concerns stem from the uniqueness of producing complex products with living recombinant hosts. It is the responsibility of each manufacturer to demonstrate to the regulatory agencies that their specific production process is adequately controlled and monitored.

1.2. Need for High and Consistent Expression of the Biopharmaceutical

A manufacturer does not really have a useable production process until it can be shown that the process yields a suitably high expression of the biopharmaceutical product (many production processes today yield expressed product at the g/L level), each time the process is run.

Also, a manufacturer does not really have an approvable production process until it can be shown that the process consistently yields a product of acceptable quality and safety.

Most production processes at Phase 1 will not meet these criteria, and they are not required to do so. Unfortunately, many biopharmaceutical companies after they rush into the clinic with a production process, fail to commit to do process development alongside their clinical development; and if they find clinical value in their pivotal clinical trials, they are then faced with a production process that either may not meet regulatory expectations or cannot be run economically. Continual development and improvement of the production process during the clinical drug development stages is essential for eventual commercial success.

1.3. What is a 'Production Process'?

The goal is to go from the gene to an expressed biopharmaceutical product. Ideally, this should occur under conditions that are stable, and conditions that produce the desired biopharmaceutical in the required amounts over a realistic timeline. And furthermore, produce the biopharmaceutical under conditions that yield a safe product and one with appropriate quality characteristics.

Regardless of the recombinant host used, the production process for a biopharmaceutical, which starts with source material (i.e., recombinant bank), consists of the following three process steps (expansion, expression and harvest), as illustrated in Figure 9.

Figure 9. Steps of a biopharmaceutical production process

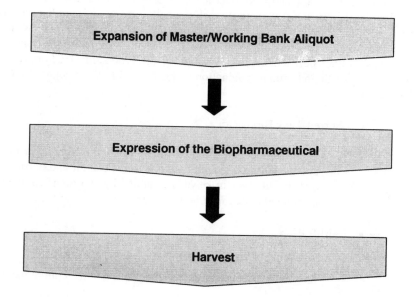

From an aliquot of the recombinant master/working bank, the source material is expanded to provide an adequate number of living organisms. For recombinant cells, this involves culturing the cells in bioreactors to obtain the necessary cell density and volume. For bioengineered animals and plants, this involves producing either enough offspring (animals) or enough seed (plants).

These expanded living recombinant organisms are then used to express the biopharmaceutical. For recombinant cells, this involves inoculating the production bioreactor with the expanded source material (commonly referred to as 'the seed culture') and controlling the bioreactor process for days or weeks, depending upon the cell type used. For bioengineered animals and plants, this involves nurturing for many months either the animals until they can lactate or crops growing in a field until they are ready to be harvested.

Once expression of the biopharmaceutical is completed, harvesting is initiated to separate the biopharmaceutical product from the recombinant host and the other components used in its production.

1.3.1. Types of Bioreactors for Cell-Based Production

The production process involves a complex set of conditions that affect cell propagation, product yield, and concentration of nutrients, waste and products. The performance of a bioreactor is governed by thermodynamics (such as the solubility of oxygen in the medium), kinetics (such as cell growth and product formation), and transport of materials (moving nutrients into the cells and removing waste products). Optimal mixing ensures effective oxygen transfer, heat transfer, and dispersal of materials. Minor deficiencies in circulation of the production medium can have major effects on growth and protein production. This has led to the design of several types of bioreactors:

- Stirred-Tank Bioreactor

    A tank that contains a motor-driven impellor or agitator to stir the cells and the culture medium, which ensures that air and nutrients are evenly distributed. Air or oxygen is added by sparging (spraying through a perforated plate in the bottom of the vessel) or surface aeration.

- Airlift Bioreactor

    Gas is sparged through the base of the tank, rising in a cylindrical draft tube to the surface, circulating the cells and the culture medium.

- Confluent Monolayer Bioreactor

    Cells are grown on matrices which provide large growth surfaces in which the cells can be grown in a confluent monolayer. The containers are rotated or agitated to keep cells in suspension. (Roller bottles are such a type of bioreactor).

- Packed Bed Bioreactor

    Cells are grown on a bed of glass or plastic beads, stainless steel bars, or a number of other materials, such as fibers and composite fabrics. A separate unit oxygenates medium that is circulated through the bed, and an aeration reservoir is used as an airlift pump.

- Hollow-Fiber Bioreactor

  Cells are grown on the outside of porous hollow fibers. Fresh medium is circulated through the fibers, allowing nutrients to diffuse through the porous fiber walls to the cells, while toxic metabolites produced by the cells diffuse into the stream flowing away.

It should be noted that some make a distinction between a fermentor (which is used for microbial cell systems) and a bioreactor (which is used for mammalian cell systems). In this CMC book, the terms 'fermentor' and 'bioreactor' will be used interchangeably; as well as the terms 'fermentation' and 'cell culturing'.

Because of the differences in characteristics between microbial and mammalian cells, the design of a bioreactor will be different for each cell type. For example, a mammalian cell requires an extended processing time compared to a microbial cell, thus requiring more stringent aseptic operating conditions. And mammalian cells are more easily damaged by shear than microbial cells, thus requiring more gentle agitation systems.

1.3.2. Harvesting Procedures for Biopharmaceuticals

For recombinant cell culture production, the expressed protein could be either intracellular or extracellular. For recovery of intracellular proteins, the cells are disrupted by chemical, enzymatic or physical methods; following disruption, cellular debris is removed by centrifugation or filtration. For recovery of extracellular proteins, the primary separation of product from cells is accomplished by centrifugation or depth filtration (works on the principles of mechanical sieving and adsorption) or tangential flow filtration (TFF; also know as crossflow filtration; works on the principle that continuously recirculated cells pass along membrane surfaces while the liquid filtrate, which contains the product, is collected).

For bioengineered animals and plants, harvesting involves either milking the animal or manually separating the seeds containing the expressed biopharmaceutical from the plant biomass.

1.4. Production Processes Familiar to the FDA

Regulatory agencies, since they review and approve the production process to grant market approval, are very familiar with production involving most types of recombinant cells. Table 28 presents some production specifics for production processes employing recombinant bacterial, yeast, mammalian or human cells.

Table 28. Production process information from FDA approved biopharmaceuticals; listed on their website: www.fda.gov/cber/products.htm

Production System	Production Process Information

Bacterial Cells

Escherichi coli
(E. coli)

Aldesleukin, Recombinant Interleukin-2 (Proleukin®)

Aldesleukin, IL-2, is produced by recombinant DNA technology using a genetically engineered *E. coli* strain containing an analog of the human interleukin-2 gene. Proleukin is not glycosylated because it is derived from *E. coli*. The manufacturing process involves fermentation in a defined medium containing tetracycline hydrochloride. The presence of the antibiotic is not detectable in the final product.

Yeast Cells

Saccharomyces cerevisiae

Recombinant Hepatitis B Vaccine (Recombivax HB®)

Recombinant hepatitis B vaccine is a non-infectious subunit viral vaccine derived from Hepatitis B surface antigen produced in yeast cells. The fermentation process involves growth of the recombinant *Saccharomyces cerevisiae* on a complex fermentation medium which consists of an extract of yeast, soy peptone, dextrose, amino acids and mineral salts. The HBsAg protein is released from the yeast cells by cell disruption and purified.

Mammalian Cells

Chinese Hamster
Ovary (CHO)

Recombinant Tenecteplase, Tissue Plasminogen Activator (TNKase®)

Tenecteplase is a tissue plasminogen activator (tPA) produced by recombinant DNA technology using an established mammalian cell line (Chinese Hamster Ovary cells). Cell culture is carried out in nutrient medium containing the antibiotic gentamicin (65 mg/L).

Baby Hamster
Kidney (BHK)

Recombinant Coagulation Factor VIIa (NovoSeven®)

Recombinant coagulation factor VIIa is produced in baby hamster kidney cells. It is secreted into the culture media in its single-chain form and then proteolytically converted to the active two-chain form during a chromatographic purification process.

Mouse Myeloma
(NS0)

Palivizumab, Humanized Monoclonal Antibody (Synagis®)

Palivizumab is produced in serum-free medium using a stirred-tank fed-batch process. A vial of cells from the WCB is expanded through a series of T-flasks and spinner flasks.

From the spinner flasks, the cells are pooled and inoculated into an expansion bioreactor. The culture is transferred into a production bioreactor for the final stage of cell culture production where glucose and nutrient feeds are supplied to the culture. The estimated numbers of cell doubling required for a commercial production process is approximately 28 population doublings from the WCB.

The bioreactors and the harvest tank were validated to be closed systems. The vessels are equipped with steam-sterilizable ports for aseptic sampling and addition of inoculum, media, and nutrient feeds. Prior to inoculating the expansion bioreactor, the spinner flask cultures are pooled into an autoclaved transfer bottle in the Class 100 BSC. The transfer bottle is equipped with a dip-tube which is connected to a silicone tubing transfer line at the bottle cap. At the end of the transfer line, a closed diaphragm valve is employed to protect the sterile envelope. The filled transfer bottle is transported on a cart to the expansion bioreactor and the diaphragm valve is connected to the inoculation port with a tri-clamp. The connection is steam sterilized for 45 ± 5 minutes. After the line cools down, the diaphragm valve and inoculation valve are opened and the culture is transferred in the expansion bioreactor using a peristaltic pump. Prior to inoculating the production bioreactor, a flexible transfer line is employed to connect the bottom drain valve of the expansion bioreactor and inoculation port of the production bioreactor. The transfer line and the tri-clamp connections are steam sterilized in place for 45 ± 5 minutes. After the line cools down, the valves are opened and the expansion bioreactor is pressurized to transfer the culture into the production bioreactor.

In the event of a bioreactor contamination, the broth is discharged into the sodium hypochlorite-based waste treatment system and the bioreactor is cleaned either manually (expansion bioreactor) or by CIP (production bioreactor) and steam-sterilized. An investigation is launched to identify the microbial contaminant(s) and the probable cause(s) of contamination. A corrective action plan is devised and implemented to eliminate the problem(s).

The cells and debris are removed from the culture using tangential-flow microfiltration, and the pH and conductivity of the resulting cell-free conditioned medium are adjusted for further processing.

Human Cells

Human Embryonal Kidney (HEK293)

Drotrecogin alfa, Recombinant Human Activated Protein C (Xigris®)

Drotrecogin alfa is a recombinant form of human Activated Protein C. An established human cell line possessing the complementary DNA for the inactive human Protein C zymogen secretes the protein into the fermentation medium. Fermentation is carried out in a nutrient medium containing the antibiotic geneticin sulphate. Geneticin sulphate is not detectable in the final product. Human Protein C is enzymatically activated by cleavage with thrombin and subsequently purified.

Since no animal or plant transgenic process has received market approval to date, the level of regulatory experience with these production processes is limited.

2. ADEQUATE DESCRIPTION OF THE PRODUCTION PROCESS

The FDA has indicated that they expect the manufacturer to make the following statement in their Phase 1 IND submission: "the sponsor believes that the manufacturing of either the drug substance or the drug product does not present any signal of potential human risk."[16]   To support such a statement, the manufacturer needs to describe the production process for the biopharmaceutical. This description needs to be accurate and of sufficient depth to allow the FDA reviewer to conclude that the production process is indeed safe and does not impose any safety risks to the expressed biopharmaceutical.

Because production processes vary with the various types of recombinant hosts that can be used, the submitted description must correctly describe the specific process being used. Recombinant organisms such as bacteria, yeast, mammalian and human cells are typically expanded in stainless steel bioreactors varying in size from 10 liters all the way up to 10,000 liters or more. Transgenic animals are raised under controlled animal husbandry practices on a ranch; while transgenic plants are cultivated under controlled agricultural practices on a farm.

The amount of information about the production process that the regulatory agencies expect to see included in the regulatory submissions increases as the clinical development advances. As stated in the FDA regulations, 21 CFR 312.23(a)(7)(i):

> "Although in each phase of the investigation sufficient information is required to be submitted to assure the proper identification, quality, purity and strength of the investigational drug, the amount of information needed to make that assurance will vary with the phase of the investigation, the dosage form, and the amount of information otherwise available."[34]

Keep in mind, that for biopharmaceuticals, the regulatory expectation is that "more information may be needed to assess the safety of biotechnology-derived drugs" compared to chemically-synthesized drugs.[16]

2.1. During Clinical Development

The minimum CMC continuum applies to production processes. At each stage of clinical development, the regulatory agencies expect that the minimum CMC requirements will be met, and that these requirements will increase and tighten as clinical development progresses.

2.1.1. Phase 1 IND Submission

For the Phase 1 IND, the amount of information needed to describe the production process is expected to be 'brief':

> "The full address of the manufacturer of the clinical trial drug substance should be submitted.

A brief description of the manufacturing process, including a list of the reagents, solvents, and catalysts used, should be submitted. A detailed flow diagram is suggested as the usual, most effective, presentation of this information."[16]

But what is 'brief'? My recommendation is that sufficient information about the production process needs to be provided to assess safety of the biopharmaceutical. Ask yourself, what would you need to know about the production process to make this assessment? That's what should be provided. Clearly, this doesn't mean submitting the batch production records. But it does mean summarizing enough of the specifics of those batch production records and including them in the submission. An example of the level of CMC detail that could be provided in the submission at this stage of clinical development is presented in Table 29. This table only presents one step of the entire production process, so similar sections would need to be written for each of the steps of expansion, expression and harvest. In addition, sections would need to be written on (1) the acceptance criteria for all of the raw materials used in the cell culture process, (2) the qualification of the bioreactor (e.g., media hold studies), (3) any specialized preparation of media, and (4) for the in-process controls to be used. In addition, the requested detailed flow diagram needs to be included.

Table 29. Illustration of the level of CMC content for a Phase 1 IND submission: Operation of the fermentor

---

Prepare the 1000 L fermentor (800 L operating volume) by calibrating probes for pH and dissolved oxygen. Attach connection for glucose, antifoam and alkali. Zero the load cell and set the operating parameters. Fill the fermentor with Water for Injection to the operating volume and then add the following fermentation raw materials (amount per L):

Yeast extract, Bacto	29.4 - 30.6 g
KH_2PO_4 (NF)	3.92 - 4.08 g
Na_2HPO_4 (USP)	4.90 - 5.10 g
......	
......	
Pluronic antifoam	added as needed for foam control (~ 50-350 mL)

Adjust pH in fermentor to 6.5 with 2.5 N NaOH. Steam sterilize medium in fermentor. Sterilize feed and addition vessels. Cool fermentor. Inoculate fermentor with 1 ampoule of the thawed Working Cell Bank, through a sterile inoculation port. The fermentation process is controlled at 37°C and pH 7.0 for about 10 to 15 hours. Airflow is kept constant at 80 LPM throughout the fermentation. The dissolved oxygen is maintained at above 20% by automatically increasing the agitation rate from 150 to 700 RPM. When the maximum 700 RPM is reached, the agitation control is switched to a manual constant 700 RPM for the rest of cultivation. Any further demand for dissolved oxygen is met by supplementary oxygen in the range of 0 to 50 LPM. When the OD_{600} is 10 to 15, a 20% yeast extract solution is fed to the fermentor at 20 g/min constant rate. When the residual glucose concentration is less than 2 g/L, a 50% (w/v) dextrose (USP) feed, under dissolved oxygen-stat control (controlled at 20-40%), is activated. When the OD_{600} reaches 30-40 (about 6 to 8 hours elapsed time), the culture is induced for recombinant protein production by adding 1.5 L of 1M tryptophan (USP) to a final concentration between 10 to 15 mM.

Four hours after induction, with an OD_{600} of 50 to 70, the fermentation culture is ready for cooling and subsequent harvest. Before harvest, several broth samples are taken aseptically for QC release tests, including non-host contamination and bacteriophage infection. When the broth is chilled to less than 20°C, the culture is harvested by centrifugation. The harvest cell paste is stored in polycarbonate tubes frozen at –70°C or below until needed.

2.1.2. Phase 2 IND Submission

It is expected that additional knowledge about the production process, especially how to more adequately control it, will be obtained during the clinical development stages. The FDA expects this additional information to be included in a Phase 2 submission, and also updates of any production process changes:

> "The general description of the synthetic and manufacturing process (e.g., ... fermentation ...) described in Phase 1 should be updated from a safety perspective if changes or modifications have been introduced.
>
> An updated detailed flow diagram for the synthesis or manufacturing process should be provided.
>
> Reagents (including solvents and catalysts), equipment (fermentors ...), and provisions for monitoring and controlling conditions used in each step should be provided.
>
> To the extent possible in Phase 2, sponsors should document that the manufacturing process is controlled at predetermined points and yields a product meeting tentative acceptance criteria. Although in-process controls may still be in development, information on in-process controls for monitoring adventitious agents should be provided."[36]

2.1.3. Phase 3 IND Submission

By the time of the Phase 3 submission, a more complete description of the production and its control is expected to be filed:

> "Updates of the acceptance criteria and analytical procedures for assessing the quality of starting materials should be provided. A table listing all reagents, solvents, catalysts should be submitted that includes (1) the grade of each

material used, (2) the specific identity test performed, (3) the minimum acceptable purity level, and (4) the step of the synthesis and manufacturing process in which it is used. For special reagents (e.g., sera, enzymes, or proteins of animal origin), a more comprehensive list of tests, screening and acceptance criteria may be needed.

An updated detailed flow diagram should be provided. Reagents, equipment (e.g., fermentors ...) and monitored and controlled conditions used in each step should be identified.

A general step-by-step description of the synthesis and manufacturing processes should be provided. Relevant information should indicate batch size (range), the type of reaction vessel, relative ratios of reactants, general operating conditions (time, temperature), in-process controls (complete description of analytical procedures) and literature references for novel reactions or complex mechanisms.

For specified biotechnology-derived products, validation of the genetic stability of the cells in production, with defined passage limits, should be performed.

Controls at selected stages in the synthesis and manufacturing process that ensure reaction completion, identity, and purity or proper cell growth should be described. The acceptance criteria and analytical procedures should be described for isolated intermediates that require control. Tentative acceptance criteria can be used to allow for flexibility in the development process, but should fulfill the primary purpose of quality control. The description of the analytical procedures should be brief, and appropriate validation information should be available on request."[36]

2.2. Preparing the BLA/NDA Submission

The 'complete' description of the production process is required in the market approval submission. And when the FDA says 'complete', they mean it. In fact, if this production information is not complete, it can be a cause for the FDA to refusal to file the submission.[31] Some of the CMC information on the production process that is required is presented in Table 30.

Table 30. CMC information to be included in the BLA/NDA dossier to describe a production process utilizing recombinant cells[40]

Identification of Manufacturer(s)	The application should include the name(s), address(es), FDA registration number, and other pertinent organizational information for each manufacturer performing any portion of the manufacture of the drug substance. A brief description of the operations performed at each location, the responsibilities conferred upon each party by the applicant and a description of how the applicant will ensure that each party fulfills their responsibilities should be submitted.
Floor Diagram(s)	For each manufacturing location, a floor diagram should be included that indicates the general facility(ies) layout. This diagram need not be a detailed engineering schematic or blueprint, but rather a simple drawing that depicts the relationship of the subject manufacturing areas, suites, or rooms to one another, and should indicate other uses made of adjacent areas that are not subject of the application. This diagram should be sufficiently clear such that the reviewer may visualize the flow of the production and would be able to identify areas or room 'proximities' that may be of concern for particular operations, e.g. segregation of pre and post viral inactivation material and operations, segregation of animal facilities, etc. Room numbers or other unique identifiers should be provided, however it is not necessary to include the location of processing equipment with rooms and areas.
Other Products	A comprehensive list of all additional products to be manufactured or manipulated in the areas used for the product should be provided. The applicant should indicate in which rooms the additional products will be introduced and the manufacturing steps that will take place in the room. An explanation should be given as to whether these additional products will be introduced on a campaign basis or concurrently during production of the product which is the subject of the application. Any additional products that may share product contact equipment with the product in question should be indicated (dedicated vs multi-use equipment should be delineated). A brief description should be provided as the type and developmental status of the additional products.
Contamination Precautions	For all areas in which operations for the preparation of cell banks and product manufacturing are performed, the following information concerning precautions taken to prevent contamination or cross-contamination should be provided: • Air quality classification of room or area in which operation is performed, as validated and measured during operations • A brief, narrative description of the procedures and/or facility design features for the control of contamination, cross- contamination and containment (air pressure cascades, segregation of operations and product, etc.) – this is of particular importance for multi-use areas or for work with live organisms • General equipment design description, e.g., does design represent and open or closed system or provide for a sterile or non-sterile operation • A description of the in-process controls performed to prevent or to identify contamination or cross-contamination The manipulation of more than one cell line in a single area, or the use of any piece of equipment for more than one cell line, should be indicated and measures to ensure prevention of cross-contamination should be discussed
Raw Materials and Reagents	A list of all components used in the manufacture of the drug substance, and their tests and specifications or reference to official compendia should be provided. For purchased raw materials representative certificates of analysis from the supplier(s) and/or manufacturer's acceptance criteria should be included in the submission. Process gases (e.g., air, carbon dioxide) and water are considered raw materials.

A list with tests and specifications of all special reagents and materials used in the manufacture of the drug substance, e.g., culture media, buffers, sera, antibiotics, etc. should be submitted.

A description of the tests and specifications for materials of human or animal source that may potentially be contaminated with adventitious agents, e.g., mycoplasma, BSE agent, and other adventitious agents of human and animal origin should be submitted. Validation data or certification supporting the freedom of reagents from adventitious agents should be included in the submission.

Process Flow
Charts

A complete visual representation of the manufacturing process flow should be provided. This flow chart should indicate the step in production, the equipment and materials used, the room or area where the operation is performed and a complete list of the in-process controls and tests performed on the product at each step. This diagram should also include information (or be accompanied by a descriptive narrative) on the methods used to transfer the product between steps, i.e., sterile, steam-in-place (SIP) connection, sanitary connection, open transfers under laminar flow units, etc. Such transfers should be described for movement of product between equipment, areas/rooms, buildings and sites. Manufacturing steps which are computer controlled should be identified.

End of
Production
Cells (EPC)

A detailed description of the characterization of the EPC that demonstrates that the biological production system is consistent during growth should be provided. The results of the analysis of the EPC for phenotypic or genotypic markers to confirm identify and purity should be included. This section should also contain the results of testing supporting the freedom of the EPC from contamination by adventitious agents. The results of restriction enzyme analysis of the gene constructs in the EPC should be submitted.

Cell Growth
and Harvesting

A detailed description of the process of inoculation, cell growth and harvesting should be submitted. The composition of the media, equipment preparation and sterilization, as well as fermentation medium sterilization, should be described. For all stages of any fermentation process the procedures which prevent contamination with adventitious agents should be described.

The stages of cell growth should be described in detail including the selection of inoculum, scale-up for propagation, and established and proposed (if different) production batch size. All operating conditions and in-process controls should also be described and appropriate ranges for operating and control parameters, such as fermentation time, cell doubling time, cell culture purity, cell viability, pH, CO_2, etc., established. If induction is required for production of protein, detailed information including induction conditions and controls employed should also be described.

The submission should include the process used to inactivate cells utilized in the production of a drug substance prior to their release into the environment. For cell lines meeting the criteria of Good Large Scale Practice (GLSP) organisms, which do not require inactivation prior to release into the environment, the information supporting their qualification as GLSP organisms should be provided. A description of the procedures used, in the event of a contamination, to inactivate a GLSP culture prior to release should be included.

If the culture supernatant or cell pellet is stored prior to processing, data supporting its stability during storage should be provided.

The manipulation of more than one cell line in a single areas or the use any piece of equipment for more than one cell line should be indicated and measures to ensure prevention of cross-contamination should be discussed.

Batch Records	A completed (executed) representative batch record of the process of production should be submitted.
In-Process Controls	A description of the methods used for in-process controls, e.g., those involved in fermentation and harvesting, should be provided.
Process Validation Studies	A description and documentation of the validation studies which identify critical parameters to be used as in-process controls, to ensure the success of routine production should be submitted.
	If the process was changed or scaled up for commercial production and involved changes in the fermentation steps, the re-validation of cell line stability during growth should be described, and the data and results provided.
	A description and documentation of the validation studies for any processes used for media sterilization, inactivating cells prior to their release to the environment, if such inactivation is required, should be provided.

3. VALIDATION OF A CELL-BASED PRODUCTION PROCESS

Validation, as defined by both the FDA and EMEA, is the means of ensuring and providing documentary evidence that the process (within its specified design parameters) is capable of consistently producing a biopharmaceutical product of the required quality.

Various manufacturers have differing opinions of the value of process validation. Some view validation as a 'curse', invented by regulators and pushed onto a company by QA. This reaction happens typically after senior management realizes the expense (typically a million dollars or more) to complete process validation for their biopharmaceutical.

The concept of process validation is a recognition that quality cannot be tested into a product. Also, it is very clear that a regulatory agency will not permit a biopharmaceutical product to be released to the market without it. The absence of process validation information in a filed BLA or NDA is sufficient reason for the FDA to issue a refusal to file (RTF):

> "The following are examples of basis for RTF due to insufficient description of the manufacturing processes: ... incomplete description of production of bulk, including description of fermentation, harvest and purification procedures, process validation ...."[31]

But even if it was not a regulatory requirement, it would be important for a manufacturer to perform process validation in order to minimize process failures, which could potentially endanger patient safety.

3.1. When Should Validation of the Production Process Occur?

Production validation should occur when the production process is defined and locked in. This clearly will not occur at Phase 1 or Phase 2. Most companies define and lock in their production process for the manufacture of Phase 3 clinical materials. Other companies wait until after the Phase 3 manufacture to do this.

At the early stages of clinical development (i.e., Phase 1 and 2), the regulatory agencies do not expect the production process to be validated:

> "Process validation for the production of APIs for use in clinical trials is normally inappropriate, where a single API batch is produced or where process changes during API development make batch replication difficult or inexact. The combination of controls, calibration, and, where appropriate, equipment qualification ensures API quality during this development phase."[22]

During clinical development is the time that the manufacturing process needs to be flexible and constantly being improved upon. However, it should be noted that because of the absence of a validated process during clinical development, there is much greater pressure placed on the Quality systems within a company:

> "The production of investigational medicinal products involves added complexity in comparison to marketed products by virtue of the lack of fixed routines, variety of clinical trial designs, consequent packaging designs, the need, often, for randomisation and blinding and increased risk of product cross-contamination and mix up. Furthermore, there may be incomplete knowledge of the potency and toxicity of the product and a lack of full process validation .... This increased complexity in manufacturing operations requires a highly effective Quality system."[49]

It is expected that during clinical development, the manufacturer is gaining sufficient information about the behaviour and physicochemical properties of the biopharmaceutical product, as well as the production process itself, to begin to clearly define the critical steps of the process and to implement appropriate in-process controls. Once the production process is locked in, it is these critical production steps and in-process controls that are then validated.

Validation studies can be performed on scaled-down processes. However, it is important that production validation be confirmed on the commercial size batches to ensure that there were no significant differences with the scale-down studies.[3] For example, operating parameters for a smaller bioreactor don't necessarily scale up to a commercial size bioreactor, and slight differences can translate into differences in the molecular variants obtained during expression of the biopharmaceutical.

3.2. Five Major Areas Involved in Validation of the Production Process

Unlike other areas of validation (such as the validation of autoclave sterilization), the validation of a production process is most difficult. To achieve validation, a production process must have a controlled method of growing the cells, must reliably express the product of interest, which can be reproducibly recovered during harvest, and consistently meet the required quality requirements.

It needs to be emphasized that one of the biggest pitfalls in process validation is that planning isn't done early enough. It can't be tacked on at the end of the development process when the BLA/NDA is under FDA review. Validation activities are costly and time-consuming. The lack of a clear understanding of the reason and the timing for validation, results in inefficient validation plans and insufficient time and resources to adequately complete validation projects.

There are 5 major areas that need to be considered under validation of production processes:

1.   The production facility, utilities and process equipment

2.   Monitoring of growth parameters

3.   In-process controls

4.   Genetic stability

5.   Cleaning validation

3.2.1. The Production Facility, Utilities and Process Equipment

The design and operation of the production facility is the first step in understanding what will be necessary for validation of the production process. The facility could be for single-product manufacturing or for multi-product campaign manufacturing or for multi-product concurrent manufacturing. Some manufacturers have even designed their production facility for multi-host concurrent manufacturing.

As the complexity of the activities in the production facility increase from single-product single-host to multi-product multi-host, the regulatory agencies have greater concern about contamination and cross-contamination issues. And they will have great interest in the manufacturer's capability of controlling this through (1) segregation of operations, (2) procedural controls, (3) sharing of process equipment, and (4) air handler design.

The FDA has made it crystal clear that the biopharmaceutical manufacturing facility, utilities and process equipment must be completely validated in order to pass the pre-approval inspection for market approval. This applies equally to small-scale pilot facilities as well as to large-scale multi-product commercial facilities.[79] This validation is accomplished by the 3 'Q's':

- Installation Qualification (IQ)

  Documented verification that the equipment or systems, as installed or modified, comply with the approved design, the manufacturer's recommendations and/or user requirements

- Operational Qualification (OQ)

  Documented verification that the equipment or systems, as installed or modified, perform as intended throughout the anticipated operating ranges

- Performance Qualification (PQ)

  Documented verification that the equipment and ancillary systems, as connected together, can perform effectively and reproducibly based on the approved process method and specifications

It should be noted that the trend today is to also consider two other 'Q's', which precede the IQ:

- Design Qualification (DQ)

  Documented verification that the proposed design of the facilities, equipment, or systems is suitable for the intended purpose

- Construction Qualification (CQ)

  Documented verification that the construction of the facilities, equipment, or systems was carried out according to the approved design

But how much of this facility, utilities and equipment validation has to be completed for the early stages of clinical development (i.e., for Phase 1 and Phase 2 clinical trial materials)? ICH Q7A, Section 19 on APIs for Use in Clinical Trials provides some insight:

"During all phases of clinical development, including the use of small-scale facilities or laboratories to manufacture batches of APIs for use in clinical trials, procedures should be in place to ensure that equipment is calibrated, clean and suitable for its intended use."[22]

So during clinical development, validation is not required, but it should be pointed out that in order to have a meaningful calibration system and to ensure that the process equipment is adequately clean, the utility systems and the equipment need to have been at least qualified (at a minimum IQ and OQ).

When it comes to safety, there are no shortcuts allowed, even for Phase 1 clinical material manufacturing. Since the bioreactor is expected to operate under 'sterile conditions' even for production of Phase 1 clinical materials, this process equipment needs to be validated for sterilization (e.g., temperature profiling and biological indicators) and for maintenance of sterility during the expected time of cell culturing (e.g.., media hold studies).

3.2.2. Monitoring of Growth Parameters

It is important for a bioreactor system to be closely monitored and tightly controlled to achieve the proper and efficient expression of the desired product. Cells deteriorate, die and disintegrate (lyse) when they get too few nutrients in the production medium. Different types of cells require different media. Many companies have switched to serum-free or protein-free media to reduce costs and avoid the issue of TSE contamination.

The parameters for the cell culture process need to be specified and monitored. There are cell culture parameters that can be directly controlled and measured on-line (i.e., input parameters):

- pH (acid/base addition)
- dissolved gases (DO_2, DCO_2)
- temperature
- pressure
- shaft speed
- component consumption rates (glucose, glutamine)
- waste byproduct build-up (ammonia, lactate)
- liquid flow rates (nutrient feeds, inducers, anti-foams)
- gas flow rate
- gas composition
- foam level

And there are cell culture parameters that are determined by sampling and then calculation (i.e., output parameters):

- absence of adventitious agents
- cell viability
- cell density
- product expression level
- impurities at harvest

It should be noted that computer programs used to control the course of cell culture growth, data logging, and data reduction and analysis need to be validated for the pre-approval inspection.[3]

Validation of the production process is an attempt to link specific values for input parameters to those for output parameters, and establish ranges for critical process parameters that will minimize variability and ensure the product will meet quality expectations. Unfortunately, there are not enough bioreactors in the world to study each parameter individually. Therefore, production validation will be accomplished by several approaches:

- Retrospective analysis of completed production runs during clinical development

- Qualified scaled-down studies to identify critical parameters and to look for important interactions, using design-of-experiment (DOE) statistical designs

- Worst-case production runs

3.2.3. In-Process Controls

In-process controls are required to ensure the absence of adventitious agent contamination and to demonstrate that the cell growth and product expression proceeded as expected.

A representative sample of the unprocessed bulk, removed from the production bioreactor prior to further processing, represents one of the most suitable levels at which the possibility of adventitious agent contamination can be determined with a high probability of detection. For recombinant bacterial and yeast cell-based production systems, microbial contamination is the many concern. For recombinant mammalian and human cell-based production systems, the concern extends to mycoplasmal and viral contamination.

To ensure acceptable product safety, the following tests should be run on each unprocessed bulk lot, beginning with the Phase 1 production runs:

- Sterility

    Sterility tests can be run with an acceptance criterion of 'no growth', but more frequently bioburden testing is performed with assigned acceptance limits.[29]

    Microbial contamination observed during the cell culturing process needs to be identified, as appropriate, and the effect of its presence on product quality should be assessed. The results of such assessments should be taken into consideration in the disposition of the material produced. Records of contamination events should be maintained.[22]

- Mycoplasma

  Mycoplasma is the simplest and smallest, self-replicating organism. Mycoplasma naturally infects arthropods, plants, and warm-blooded animals. In humans, mycoplasma is most commonly associated with pneumonia. Production culture contamination usually originates from components of cell culture medium, such as serum, or from an infected person working in production.

  Testing for mycoplasma must include cultivable and non-cultivable species.[80]

  Unfortunately, the FDA required mycoplasma assay takes 28 days to complete with several more weeks before the QA-audited report is completed. This is a concern to manufacturers because once mycoplasma contamination is present in the bioreactors it can become a serious problem for subsequent runs. Several manufacturers have reported mycoplasma contaminations in their large-scale production bioreactors[81, 82] For this reason, many companies now employ assays using the polymerase chain reaction (PCR) to rapidly test for mycoplasma in the production culture, as an early warning system. These PCR mycoplasma DNA tests can be readily performed within a work day, allowing manufacturing and the Quality Unit the opportunity to assess the condition of the bioreactor culture prior to breaking the bioreactor containment.

  If mycoplasma contamination is detected, the production lot should not be used or processed further.[29]

- Virus

  A representative sample of the unprocessed bulk, removed from the production reactor prior to further processing, represents one of the most suitable levels at which the possibility of adventitious virus contamination can be determined with a high probability of detection. It is recommended that manufacturers develop programs for the ongoing assessment of adventitious viruses in production batches.[69]

  In vitro virus testing with at least three indicator cell lines (e.g., Vero, MRC5, 3T3) should be performed routinely. When contamination with a particular virus is encountered in a facility, consideration should be given to modifying the routine testing program in order to detect that virus.[29]

In vivo virus testing is generally done once (as part of the master bank characterization) but should be repeated when production methods change.[29]

Retrovirus contamination should be quantitated on three clinical grade production lots in order to establish the level of virus contamination for the specific cell line and manufacturing process. This quantitation of retrovirus should be done preferably by generic assays such as transmission electron microscopy (TEM) or alternatively by sensitive infectivity assays. Quantitation should be repeated when changes in culture media, duration or scale of production are made.[27]

Bioreactors containing hamster cells (recombinant CHO) can become contaminated with murine minute virus (MMV; also known as minute virus of mice, MVM) that may escape detection in the routine in vitro assays. Since these in vitro assays take nearly a month to complete, this presents a major manufacturing risk. Once a virus is present in the bioreators it can become a serious problem for subsequent runs. One manufacturer has encountered this with a MMV contamination in their production process at the 12,000 liter scale.[83] For this reason, many companies now employ polymerase chain reaction (PCR) tests to rapidly test for MMV in the CHO production culture prior to breaking the bioreactor containment, as an early warning system. These PCR MMV DNA tests can be readily performed within a work day, allowing manufacturing and the Quality Unit the opportunity to assess the condition of the bioreactor culture prior to breaking the bioreactor containment.

Harvest material in which adventitious virus has been detected should not be used to manufacture the biopharmaceutical product.[69]

Some additional tests that should be run on each bioreactor production lot include the following:

- Cell Growth

    During clinical development, data on the consistency of either cell density and/or cell viability over the course of the production cell growth are obtained. These growth profiles can then be used to establish a 'typical growth profile' to be used as a reference for future production runs. For the BLA/NDA filing, some manufacturers use either the typical growth profile and/or a lower limit for cell viability as acceptance criteria.

- Expression Level

Manufacturers desire to use 'super-cell lines' in production, in which efficiency of the expression system is maximized. Companies that invest in this process development may be able to downsize the size of their commercial facilities.

Data on the consistency of the expression level of the biopharmaceutical product from each bioreactor production run needs to be obtained. Criteria for the rejection of culture lots should be established.[2] An acceptance criteria range needs to be set. Lower than expected product expression in the production solution at harvest can cause an overload of impurity onto the downstream purification process. Higher than expected product expression in the production solution at harvest can cause a product overload of the downstream purification process, which could in turn cause some process steps to exceed their validated limits.

During clinical development, the product expression level is measured, but for the BLA/NDA filing, an acceptance range will have to be set.

• Endotoxin Impurity

The presence of bacterial endotoxin in the harvested production solution could be the result of a low level bacterial contamination in the bioreactor or present due to contamination of a raw material used in the production culture or due to handling of the culture solution during harvesting. Endotoxin is measured at the harvest to ensure that the downstream purification process is not overloaded with this contaminant.
During clinical development, the endotoxin impurity level is measured, but for the BLA/NDA filing, an upper acceptance limit should be established.

3.2.4. Genetic Stability

By Phase 3 clinical development, the validation of the genetic stability of the production cells is expected to be completed:

"For specified biotechnology-derived products, validation of the genetic stability of the cells in production, with defined passage limits, should be performed."[36]

The results of the genetic stability validation are used to determine the maximum 'in vitro cell age' for the production process. The in vitro cell age is the measure of time between thaw of the MCB vial(s) to harvest of the production vessel measured by elapsed chronological time, by population doubling level of the cells, or by passage of the cells when subcultivated by a defined procedure for dilution of the culture.[65] An acceptance criteria for the production process is the requirement that the in vitro cell age (i.e., the end-of-production cells in the bioreactor) not exceed the calculated maximum cell age value.

The regulatory guidance on how to carry out this validation is as follows:

"For the evaluation of stability during cultivation for production, at least two time points should be examined, one using cells which have received a minimal number of subcultivations, and another using cells at or beyond the limit of in vitro cell age for production use described in the marketing application. The limit of in vitro cell age for production use should be based on data derived from production cells expanded under pilot plant scale or commercial scale conditions to the proposed limit of in vitro cell age for production use or beyond. Generally, the production cells are obtained by expansion of cells from the WCB; cells from the MCB could be used with appropriate justification. This demonstration of cell substrate stability is commonly performed once for each product marketing application."[65]

"Evaluation of the cell substrate with respect to the consistent production of the intended product of interest should be the primary subject of concern. The type of testing and test article(s) used for such assessments will depend on the nature of the cell substrate, the cultivation methods, and the product. For cell lines containing recombinant DNA expression constructs, consistency of the coding sequence of the expression construct should be verified in cells cultivated to the limit of in vitro cell age for production use or beyond by either nucleic acid testing or product analysis...."[65]

"In the case of tissue culture or fermenter production, end-of-production cells (EPC) should be tested at least once to evaluate whether new contaminants are introduced or induced by the growth conditions."[29]

It needs to be pointed out that changes in the production process (e.g.., changes in media, scale-up, etc.) can impact the maximum in vitro cell age assignment; thus, requiring the need to re-validate this parameter (or at a minimum prepare a documented justification of why the validation is not impacted by the change).[29]

3.2.5. Cleaning Validation

Cleaning validation is a key activity in a biopharmaceutical manufacturing facility, especially if the production facility is to be operated for multi-products. Each company must determine its own cleaning policies and acceptance criteria, based upon knowledge of the equipment and processes employed. Each stage of the production process in which the equipment is being used, the maximum holding time of soiled equipment prior to cleaning, and the nature of the starting materials, intermediates and products, should be considered. Finally, the overall impact of potential residues on the safety, efficacy and purity of the product must be assessed.

The cleaning of the production process equipment (e.g., seed bioreactors, the production bioreactor, hard-piped transfer lines, etc.) must be completely validated at the time of the pre-approval inspection:

> "Cleaning procedures should normally be validated. Validation of cleaning procedures should reflect actual equipment usage patterns.
>
> Shared (multi-product) equipment may warrant additional testing after cleaning between product campaigns, as appropriate, to minimize the risk of cross-contamination."[22]

The amount of effort to complete this cleaning validation is extensive. Some elements to be designed into the cleaning validation program are presented in Table 31.

Table 31. Elements of a cleaning validation program for production process equipment, such as a bioreactor[84, 85]

Evaluation of Cleanliness	Analysis of final rinse water samples after cleaning
	Analysis of surface swab samples after cleaning
	Visual inspection of surfaces after cleaning
Test Methods	Total Organic Carbon Assay: to demonstrate removal of organic material from cells, product, media components, and cleaning agents
	Product-Specific Assay: to demonstrate removal of product
	pH: to demonstrate removal of caustic used in cleaning
	Visual Inspection: to detect visible residues (e.g., rouging) on surfaces
Acceptance Criteria	Criteria should be scientifically justifiable, practical and achievable
	Limit of assay detection or residue carryover at < 1/1000th of normal therapeutic dose
FDA's Expectations	FDA expects firms to have written procedures (SOPs) detailing the cleaning processes used for various pieces of equipment
	FDA expects firms to have written general procedures on how cleaning processes will be validated

FDA expects firms to prepare specific written validation protocols in advance of the study being performed

FDA expects a final validation report which is approved by management and which states whether or not the cleaning process is valid and that supports a conclusion that residues have been reduced to an "acceptable level"

Under no circumstances should stagnant water be allowed to remain in equipment subsequent to cleaning operations

The specificity and sensitivity of the test methods used to detect residuals or contaminants must be known; the % recovery of residuals from surfaces should be known

"Test Until Clean" (clean equipment, sample, test, further clean equipment, resample, and retest until an acceptable residue level is attained) is not acceptable, except in rare cases

But is it necessary to have cleaning validation of the production equipment at early clinical manufacturing stages? ICH Q7A, Section 19 on APIs for Use in Clinical Trials, states that process equipment needs to be clean even at Phase 1 manufacturing:

"During all phases of clinical development, including the use of small-scale facilities or laboratories to manufacture batches of APIs for use in clinical trials, procedures should be in place to ensure that equipment is calibrated, clean and suitable for its intended use."[22]

"Procedures for the use of facilities should ensure that materials are handled in a manner that minimizes the risk of contamination and cross-contamination."[22]

But does 'clean' mean 'validated to be clean' at this stage of development? For the most part, probably not. At Phase 1, there should be an established cleaning procedure and there should be some testing to ensure that the equipment is cleaned by the procedure. However, if the production equipment is product contacting and not product-dedicated, I would recommend that cleaning validation be performed on the equipment. Patients on clinical studies deserve to be protected from cross-contamination as much as those who will use the drug if it gets approved for market.

Because of the expense and resource commitments needed to complete all of the production cleaning validation studies (whether the cleaning is by manual means or clean-in-place automated means), single-use, disposable equipment is being evaluated by the biopharmaceutical industry. Even single-use, disposable bioreactors are under consideration as a cost-effective way to eliminate much of the risk of cross-contamination and the major cost of completing cleaning validation. However, they are still of limited capacity (currently in the hundreds of liters working volume) which works fine for gene therapy vector production but not so for most cell-based recombinant protein production.

3.3. Final Comments on Process Validation

Production validation is not an experiment. It is too expensive to perform over and over again at the production scale because the pre-defined acceptance criteria were not based on scientific data. It is far better to perform pre-validation experiments to gather the data necessary to set proper acceptance criteria.

Avoid information gathering for no particular purpose. Clearly understand and communicate why a specific validation protocol is necessary.

Plan adequately. It will always cost more and take longer than budgeted.

4. ADDITIONAL PRODUCTION CONTROLS AND CONCERNS

For each type of production system (the cell-based production systems, the gene therapy virus vector systems and the bioengineered animal and plant production systems), the regulatory agencies have indicated the need for including selected controls into the production process, and they have expressed additional concerns.

4.1. Cell-Based Production Processes

Describing a typical cellular production process is a challenge as there is so much diversity. There are different types of recombinant cells that can be utilized. The facilities and quality systems applied to these production process can be designed for either single-product or multi-product manufacturing. Furthermore, the facilities and quality systems can be designed for manufacture of either only early stage clinical trial materials (i.e., Phase 1 and 2 clinical trial materials) or late stage clinical trial materials (i.e., Phase 3 clinical trial materials) and commercial product. The scale of the production process equipment can be designed to handle bioreactor sizes of either 10 liters, 100 liters, 1000 liters, 10,000 liters or larger.

Regardless of the production facility or the process design or the scale of the production process, there are 4 additional key elements involved in the control of a cellular production process for any biopharmaceutical:

1. Cell culture media acceptance criteria

2. Avoidance of animal- and human-derived raw materials

3. Containment of the recombinant organism

4. Contamination control for aseptic processing operations

4.1.1. Cell Culture Media Acceptance Criteria

Cell culture media must be carefully selected to provide the proper rate of growth and the essential nutrients for the recombinant organisms producing the biopharmaceutical. However, this media should not contain any undesirable and toxic components that may be carried through the production process into the purification process and into the finished product.

Production media designed for high-density cell cultures contain high levels of glucose and glutamine because these constituents are the primary source of metabolic energy for cell density, longevity and expression. However, high levels of glucose can result in high levels of lactate through glycolysis. Lactate accumulation can reduce the pH throughout the culture, and that low pH can be detrimental to cell viability and productivity. This problem can be addressed by adding a base to the culture; however, neutralizing metabolic acids can increase osmolality, which if increased too rapidly and too high can negatively affect growth and protein expression, limiting bioreactor longevity. Furthermore, high levels of glutamine can result in high levels of ammonia through either metabolic hydrolysis to glutamic acid or from spontaneous deamidation as a result of medium storage. Elevated ammonia levels can lead to reduced expression.

Many culture media are proprietary to the vendor, so there is a reluctance to provide this information to the biopharmaceutical manufacturer. However, the manufacturer needs to negotiate making this information available to its Quality Unit and to the FDA.

Water is the major raw material in the production media. The quality of the water that a manufacturer will produce depends on the recombinant system used, with many manufacturers using Water for Injection (WFI) in the production process. Because the water generation systems in biopharmaceutical facilities are complex (consisting of water treatment units, distillation units, hot and cold distribution loops and terminal filters at use points), continual testing of water used in the production process per the United States Pharmacopeia (USP) standard is required.[86]

Penicillin and other beta lactam antibiotics should not be present in production cell cultures. Minimal concentrations of other antibiotics (such as tetracycline, neomycin and gentamicin) may be acceptable in the production cell cultures.[80]

4.1.2. Avoidance of Animal- and Human-Derived Raw Materials

Animal- and human-derived raw materials in production media should be avoided if at all possible because of the risk of viruses and TSEs. However, many mammalian cell culture media still contain these animal-derived materials (e.g., fetal calf serum, casein, animal hydrolysates, cholesterol, trypsin) and human-derived materials (e.g., human transferring, human plasma protein solution), primarily because they significantly impact the productivity of the cells. In these cases, companies run a significant risk if they only rely on a vendor's certificate of analysis for these materials. It is good business to perform a quality assurance audit of the vendor and its testing lab. Some companies also independently assay incoming lots of these raw materials to confirm a vendor's virus results. Ask for the regulatory package from the vendor on these specific materials. In addition, check if the vendor has an European Directorate for the Quality of Medicine (EDQM) certificate of suitability (CEP) for their specific raw material that meets the European Pharmacopoeia (PHEUR) requirements for control and evaluation of TSE risk.[87]

Companies should be well advised of the potential risks and regulatory hurdles for using the animal- and human-derived raw materials in their production processes. As discussed previously under minimizing the risk of TSEs for master/working banks, the FDA has issued a strong recommendation to the industry:

> "CBER strongly recommends that manufacturers take whatever steps are necessary to assure that materials derived from all specifies of ruminant animals born, raised or slaughtered in countries where BSE is known to exist, or countries where the USDA has been unable to assure FDA that BSE does not exist, are not used in the manufacture of FDA-regulated products intended for administration to humans."[75]

In Europe, the EMEA has issued its recommendations to the industry. Table 32 presents details on how to minimize the risk of TSE contamination. However, the EMEA recommendation goes a step farther than that made by the FDA. For the EMEA, a manufacturer not only needs to ensure that the animal-derived or human-derived materials are safe but also needs to provide a justification for why they really need to use them:

> "When manufacturers have a choice to use ruminant or non-ruminant material, the use of non-ruminant material is preferred. Substitution of ruminant source materials by material from other species which are recognized to suffer from TSEs, or which can be infected experimentally by the oral route, would not normally be acceptable.
>
> Pharmaceutical manufacturers and producers of medicinal products of animal origin are responsible for the selection and justification of adequate measures. The state of science and technology must be taken into consideration."[77]

The use of alternative cell culture media has increased significantly: serum-free media (contains no animal serum), animal-free media (contains no animal-derived components) or chemically-defined media (contains low amounts of non-animal-derived components).

Table 32. EMEA recommendations on minimizing TSE contamination in animal-derived manufacturing materials[77, 88]

Specified Risk Materials	The following specified risk materials are to be removed and destroyed: (1) the skull including the brain and eyes, the tonsils, the vertebral column excluding the vertebrae of the tail and the transverse processes of the lumbar vertebrae, but including dorsal root ganglia and spinal cord, of bovine animals aged over 12 months, and the intestines from the duodenum to the rectum and the mesentery of bovine animals of all ages (2) the skull including the brain and eyes, the tonsils and the spinal cord of ovine and caprine animals aged over 12 months or which have a permanent incisor erupted through the gum, and the spleen of ovine and caprine animals of all ages.
Selection of Source Materials	The most satisfactory source of materials is from countries which have not reported cases of BSE and have (a) a compulsory notification and (b) compulsory clinical and laboratory verification of suspected cases. In addition, it should be ensured that a risk of BSE infection is not introduced from the following factors: (a) importation of cattle from countries where high incidence of BSE has occurred, (b) importation of progeny of affected females, and (c) the use in ruminant feed of Meat and Bone Meal containing any ruminant protein which originates from countries with a high or low incidence of BSE, or of unknown status.
Levels of Infectivity in Animal Parts	Category I: High infectivity brain, spinal cord, (eye) Category II: Medium infectivity ileum, lymph nodes, proximal colon, spleen, tonsil, (dura matter, pineal gland, placenta), cerebrospinal fluid, pituitary, adrenal Category III: Low infectivity distal colon, nasal mucosa, peripheral nerves, bone marrow, liver, lung, pancreas, thymus Category IV: No detectable infectivity blood clot, faeces, heart, kidney, mammary gland, milk, ovary, saliva, salivary gland, seminal vesicle, serum, skeletal muscle, testis, thyroid, uterus, foetal tissue, (bile, bone, cartilaginous tissue, connective tissue, hair, skin, urine)
Acceptability of Use	The acceptability of a particular medicinal product containing these materials, or which as a result of manufacture could contain these materials, will be influenced by a number of factors, including: • documented and recorded source of animals • nature of animal tissue used in manufacture • production process(es) • route of administration • quantity of tissue used in the medicinal products • maximum therapeutic dosage (daily dosage and duration of treatment) • intended use of the product

According to the USDA in 2003, there are now 35 countries/areas listed as 'at risk for BSE'. Also, there are now only 3 countries/areas in the world remaining that are still considered 'scrapie-free' (Australia, New Zealand, Trust Territories of the Pacific Islands).[89]

If all of this isn't concern enough, in 1999, there was a scare in Europe when polychlorinated biphenyl (PCB) contaminated feed was fed to cows, sheep and goats. The FDA had to issue an advisory to industry about their possible presence in animal-derived manufacturing materials.[90]

The prudent step today, is not to use animal- or human-derived raw materials in the production process, if at all possible. And if these components are currently present in a production process, consider if it is at all possible to remove or replace them.

4.1.3. Containment of the Recombinant Organism

Bioreactor systems designed for recombinant microorganisms require not only that a pure culture is maintained, but also that the culture be contained within the systems. The containment can be achieved by the proper choice of a host-vector system that is less capable of surviving outside a laboratory environment and by physical means, when this is considered necessary. Good Large-Scale Practice (GLSP) level of physical containment is recommended for large-scale production involving viable, non-pathogenic and nontoxigenic recombinant strains derived from host organisms that have an extended history or safe large-scale use.[3]

4.1.4. Contamination Control for Aseptic Processing Operations

Controls and procedures must be in place, even at the start of Phase 1 product manufacturing, to ensure that the facility and equipment are handled in a manner that minimizes the risk of microbial contamination for the biopharmaceutical product.

While bioreactors are designed to provide a protective barrier around the cell culture, a single contamination event can ruin a whole production run and waste all the raw materials, labor and equipment time invested, not to mention the risk of contaminating the downstream purification process. Keeping these cell cultures free of contaminants is one of the primary objectives in the design and operation of a bioreactor.

Bioreactors come in many sizes (from 1 liter to over 10,000 liters in operating volumes) and can be made of stainless steel (the material of choice) or glass. It is most important that the expansion of the seed and the expression of the product in these bioreactors remain free of adventitious microbial agents. Therefore, preventing contamination is much more than just sterilizing the bioreactor; it also involves correct operating procedure. The production medium used in the bioreactor must be sterilized, which can be done either in the bioreactor while it is undergoing sterilization in place (SIP) or by sterile filtration of the media as it is being added to the sterile bioreactor. Any nutrients or chemicals added beyond this point also must be sterile, typically by sterile filtration into the bioreactor. Air lines to the bioreactor must include sterilizing filters. Exhaust gas lines to the bioreactor must also have sterilizing filters. The bioreactor inoculation, transfer, and harvesting operations must be done using validated aseptic techniques. Additions or withdrawals need to be done through steam sterilized lines or steam-lock assemblies.

Because of the importance of maintaining the conditions inside the bioreactor as 'sterile', even for Phase 1 clinical material production, the bioreactor should be validated both to be sterile at the start of the process and to hold sterility over the course of production.

The challenge then is to operate the bioreactors and carry out the production process without introducing adventitious contamination. According to ICH Q7A Section 18 Specific Guidance for APIs Manufactured by Cell Culture/Fermentation, the following general principles need to be applied:

> "Where cell substrates, media, buffers, and gases are to be added under aseptic conditions, closed or contained systems should be used where possible. If the inoculation of the initial vessel or subsequent transfers or additions (media, buffers) are performed in open vessels, there should be controls and procedures in place to minimize the risk of contamination.

> Where the quality of the API can be affected by microbial contamination, manipulations using open vessels should be performed in a biosafety cabinet or similarly controlled environment.

> Personnel should be appropriately gowned and take special precautions handling the cultures.

> Harvesting steps, either to remove cells or cellular components or to collect cellular components after disruption should be performed in equipment and areas designated to minimize the risk of contamination."[22]

What about the environment in the production facility, and the contribution of its 'cleanliness' on preventing adventitious contamination during production operations? For cell culture production processes, the FDA has not clearly defined their expectations for environmental conditions in a guidance document. However, in a published paper, FDA reviewers elaborated on their expectations for environmental conditions for the cell culture manufacturing areas for biopharmaceuticals:

> "Propagation of the host system comprises the initial steps in manufacturing. The product is usually particularly sensitive to microbial or cross-contamination at this stage. Therefore, open manipulations should be performed under a Class 100 hood or other local Class 100 environment whenever possible. In addition, the manufacturing areas surrounding these Class 100 areas should be classified at 100,000. In general, gowning areas used to enter classified areas should be designed with appropriate levels of control (for example, a Class 100,000 gowning room should be used before entering a Class 10,000 manufacturing area).

We generally recommend that scale-up, fermentation and harvest procedures be performed in Class 100,000 areas. For manufacturing areas containing validated closed systems, the limits may exceed those recommended for Class 100,000 environments, but these areas should continue to be environmentally monitored. Additionally, action limits should be established based on historical data and the nature of the operations that occur in the area. We recommend Class 100 conditions for any inoculations, additions, or sampling procedures for which the connections cannot be appropriately sterilized prior to transfer. Some manufacturers have achieved Class 100 conditions for these operations through the use of portable laminar flow hoods."[91]

For clarification, cleanroom classifications are based on limits for nonviable and viable airborne particulate monitoring, microbial surface monitoring, and personnel monitoring; Class 100 is also referred to as Grade A or Class M3.5 or ISO14644-1 5; Class 100,000 is also referred to as Grade D or Class M6.5 or ISO14644-1 8.[92, 93]

These cleanroom classifications need to be in place even for manufacture of Phase 1 material, since the FDA will want to review this information in the IND submission. However, assignment of the action and alert limits in the environmental monitoring program can be loose. As experience is gained with the manufacturing facility and the cell culture process, the limits can be appropriately tightened using this historical data. The final alert and action limits will need to be set, and justified, for the BLA/NDA submission.

4.2. Gene Therapy Production Processes

Gene therapy production processes either involve the cell-based production of genetically engineered plasmids or viral vectors or involve the expansion of genetically modified cells to sufficient numbers for re-administration to a patient.

4.2.1. Control of the Cells

In gene therapy production processes, cells are used, so the production controls that are used to control other cellular production processes apply here also. However, because these biopharmaceuticals are often complex mixtures that cannot be completely defined (e.g., genetically modified patient's cells), the FDA has extra concern for the control of the production process, especially in steps to ensure no introduction of adventitious agents or other contaminants. (Table 33).

The EMEA has a similar set of production controls for the gene therapy processes.[94]

Table 33. CMC controls for a gene therapy production process[35]

Raw Material Acceptance Criteria	Acceptance criteria should be established for all media and components, including validation of serum additives and growth factors, as well as verification of freedom from adventitious agents. Records should be kept detailing the components used in the culture media, including their sources and lot numbers. Medium components which have the potential to cause sensitization, for example certain animal sera, selected proteins, and blood group substances, should be avoided. For growth factors, measures of identity, purity, and potency should be established to assure the reproducibility of cell culture characteristics. Penicillin and other beta-lactam antibiotics should be avoided during production, due to the risk of serious hypersensitivity reactions to patients. If antibiotic selection is used during the production of plasmid vector gene therapy products, it is preferable not to use selection markers which confer resistance to antibiotics in significant clinical use, in order to avoid unnecessary risk of spread of antibiotic resistance traits to environmental microbes. Materials used during in vitro manipulation procedures, for example antibodies, cytokines, serum, protein A, toxins, antibiotics, other chemicals, or solid supports such as beads can effect the safety, purity, and potency of the product. These components should be clearly identified and a qualification program with set specifications should be established for each component to determine its acceptability for use during the manufacturing process. When using reagent grade material, the qualification program should include testing for safety, purity, and potency of the component where appropriate. Abbreviated testing may be appropriate for use of clinical grade components. The country of origin should be certified when there is risk of transmissible agents causing spongiform encephalopathy.
In-Process Controls	Cell Identity and Heterogeneity Both manufacturing and testing procedures should be implemented which ensure the control of cell cultures with regard to identity and heterogeneity. Cell culturing practices and facilities should be designed to avoid contamination of one cell culture with another. During cell culturing, extensive drift in the properties of a cell population, or overgrowth by a difference cell type originally present in low numbers, may occur. To detect such changes, cell identity should be assessed quantitatively, for example, by monitoring cell surface antigens or biochemical markers. The method of identification chosen should also be able to detect contamination or replacement by other cells in use in the facility. Acceptable limits for culture composition should be defined. Quantitative assays of functional potency may sometimes provide a method for population phenotyping. The desired function should be monitored when the cells are subjected to manipulation, and the tests carried out periodically to assure that the desired trait is retained. Identity testing should in some cases include verification of donor-recipient matching and immunological phenotyping. Characterization of Therapeutic Entity If the intended therapeutic effect is based on a particular molecular species synthesized by the cells, enough structural and biological information should be provided to show that an appropriate and biologically active form is present. Culture Longevity The essential characteristics of the cultured cell population (phenotypic markers such as cell surface antigens, functional properties, activity in bioassays, as appropriate) should be defined, and the stability of these characteristics established with respect to time in culture. This profile should be used to define the limits of the culture period.

Replication Competent Virus (RCV)

When either viral vectors are under production or cells have been genetically modified by viral vectors, a sensitive assay for replication competent retrovirus (RCR) or replication competent adenovirus (RCA), respectively, is required. Since the possibility of recombination can occur at any point in the production process, each production lot is to be tested. The testing is to involve 1% of the production or patient's transduced cells or 10^8 cells, whichever is fewer.

Contamination Control

Poor control of production processes can lead to the introduction of adventitious agents or other contaminations, or to inadvertent changes in the properties or stability of the biological product that may not be detectable in final product testing.

Documentation should be provided that cells are handled, propagated, and subjected to laboratory procedures under conditions designed to minimize contamination with adventitious agents. During long term culturing, cells should be tested periodically for contamination. Testing should ensure that cells are free of bacteria, yeast, mold, mycoplasma, and adventitious viruses.

4.2.2. RAC Review and Approval of the Production Process

One additional feature of gene therapy is that these production processes are also reviewed and approved by the Recombinant DNA Advisory Committee (RAC) of the National Institutes of Health (NIH). While these public RAC meetings are not required if the clinical site and sponsor for the study are entirely publicly funded, manufacturers submit their protocols to this body anyway. Table 34 presents some of the CMC information required by RAC. If a new gene therapy vector is under consideration or a new gene delivery system is being proposed, there typically is public discussion.

Table 34. CMC production information requested by RAC[95]

Appendix M-II-B-1. Structure and Characteristics of the Biological System

Provide a full description of the methods and reagents to be employed for gene delivery and the rationale for their use. The following are specific points to be addressed:

Appendix M-II-B-1-a. What is the structure of the cloned DNA that will be used?

Appendix M-II-B-1-a-(1). Describe the gene (genomic or cDNA), the bacterial plasmid or phage vector, and the delivery vector (if any). Provide complete nucleotide sequence analysis or a detailed restriction enzyme map of the total construct.

Appendix M-II-B-1-a-(2). What regulatory elements does the construct contain (e.g., promoters, enhancers, polyadenylation sites, replication origins, etc.)? From what source are these elements derived? Summarize what is currently known about the regulatory character of each element.

Appendix M-II-B-1-a-(3). Describe the steps used to derive the DNA construct.

Appendix M-II-B-1-b. What is the structure of the material that will be administered to the research participant?

Appendix M-II-B-1-b-(1). Describe the preparation, structure, and composition of the materials that will be given to the human research subject or used to treat the subject's cells: (i) If DNA, what is the purity (both in terms of being a single DNA species and in terms of other contaminants)? What tests have been used and what is the sensitivity of the tests? (ii) If a virus, how is it prepared from the DNA construct? In what cell is the virus grown (any special features)? What medium and serum are used? How is the virus purified? What is its structure and purity? What steps are being taken (and assays used with their sensitivity) to detect and eliminate any contaminating materials (for example, VL30 RNA, other nucleic acids, or proteins) or contaminating viruses (both replication-competent or replication-defective) or other organisms in the cells or serum used for preparation of the virus stock including any contaminants that may have biological effects? (iii) If co-cultivation is employed, what kinds of cells are being used for co-cultivation? What steps are being taken (and assays used with their sensitivity) to detect and eliminate any contaminating materials? Specifically, what tests are being conducted to assess the material to be returned to the subject for the presence of live or killed donor cells or other non-vector materials (for example, VL30 sequences) originating from those cells? (iv) If methods other than those covered by Appendices M-II-B-1 through M-II-B-3, Research Design, Anticipated Risks and Benefits, are used to introduce new genetic information into target cells, what steps are being taken to detect and eliminate any contaminating materials? What are possible sources of contamination? What is the sensitivity of tests used to monitor contamination?

Appendix M-II-B-1-b-(2). Describe any other material to be used in preparation of the material to be administered to the human research subject. For example, if a viral vector is proposed, what is the nature of the helper virus or cell line? If carrier particles are to be used, what is the nature of these?

4.3. Transgenic Animal Production Processes

The use of transgenic animals for production of biopharmaceuticals adds a new layer of quality and regulatory controls not needed in cell-based production systems. A transgenic animal has a limited lifetime (possibly less than 10 years). Each animal experiences its own physiological changes and various environments throughout its life (e.g., nursery, general housing, milking area, etc.) as it develops, gives birth and ultimately lactates.

4.3.1. Production Controls

Some of the CMC production controls recommended by the FDA for these bioengineered animals are presented in Table 35.

Table 35. CMC controls for a transgenic animal production process[66]

Generation of the Production Herd	Each production animal must be traceable to a particular founder animal. Also the place and date of birth, use in production, incidence and course of disease and final disposition should be recorded for each production animal.
	Any use of procedures of artificial insemination, embryo transfer, or semen collection and storage should be documented and appropriate standards applied
	Manufacturers should demonstrate that the animals to be used as recipients of transgenic sperm or embryos are healthy and are free of relevant infectious agents.
	Pregnancy should be monitored.
Selection of the Production Herd	Criteria for admission to the production herd are important for two reasons: 1) to assure that the quality and levels of the transgene product are acceptable and, 2) to prevent the introduction of an infectious agent into the herd.
	Specifications for the range of acceptable final yields should be established before considering whether the contribution of the animal is to be used directly or pooled so that the concentration of the active component in the material to be purified will be high enough to assure an adequate purification.
	To protect against adventitious agents, sick animals should never be added to the herd, and healthy animals should have met the requirements for entry into the breeding herd and been monitored for a sufficient time period.
Monitoring the Health of the Herd	Monitoring programs are important for both maintaining the health of the animals and for preventing product contamination with adventitious agents, pesticides, and animal medications.
	A detailed plan describing the monitoring plans should be available. Health records should contain the complete history from birth to death of all animals used in production, including the drugs and vaccines used. Disease episodes should be definitively diagnosed to the extent possible. Sick animals should be removed from production. The plan should include monitoring techniques, endpoints, and methods of reporting results. Methods for recording veterinary care or other preventive measures (e.g., vaccines, vitamin supplements, nutritional additives) should be established.
	An animal that grazes is not in a controlled environment and therefore may require more rigorous and extensive testing for infectious agents than those raised under barrier conditions.
	Deleterious effects may arise from over expression or insertional effects. The transgenic animal should be observed for sufficient time to establish effects of constitutive expression of a given product.
Feeding of Animals	Transgenic animals should not receive feeds which contain rendered materials derived from species which may contain TSE agents.
	Consideration should be given to monitoring feed for pesticide residues. Records of the composition of the feed of the animals and the level of consumption should be kept as part of the animal maintenance record. A change in feed consumption is frequently an early indication of disease in agricultural animals.

Housing Facilities	The containment and confinement practices for production operations involving transgenic animals should be in accordance with applicable portions of the NIH guidelines for Research Involving Recombinant DNA Molecules.
	The physical surroundings where the transgenic animals will be maintained should meet the requirements of 21 CFR 600.11.
	If the facility is not a single-species-dedicated breeding and maintenance facility, the adventitious agents of the other species must be considered.
	The surroundings should be capable of containing the animals and of preventing the accidental entry of other animals.
	Transgenic animals should be neutered after breeding to lessen the chance of escape or inadvertent breeding into the nontransgenic population.
Removal of Animals from Production	Criteria that will result in the temporary or permanent removal of animals from production need to be established. This could include illness, cessation of production, the appearance of an adventitious agent (even without clinical signs), injury or other conditions.
	Regardless of the reason for temporary removal, stringent criteria for readmission to the production herd should be established.
Animal Sentinel Program	When feasible, a sentinel animal program that will allow periodic health evaluations should be considered. Such sentinel animals should be infertile, of the same species or origin, and should be maintained with the transgenic production herd.

4.3.2. Protecting the Gene Pool

To prevent the compromise of food supplies, controls are also required to ensure that retired or dead transgenic animals are not used as food for humans or animals. This was recently emphasised to those involved in the bioengineering of animals by the FDA:

"To date, FDA has not permitted genetically engineered animals to be placed into the human food supply. Likewise, only in certain circumstances has the FDA allowed animals from genetic engineering investigations to be rendered and incorporated into animal feed."[96]

4.4. Transgenic Plant Production Processes

There are many challenges in moving biopharmaceutical production outdoors to the field. Current agricultural practices are geared toward production of commodity foods in large volumes, but these practices are insufficient when it comes to biopharmaceutical production. Grain production starts with soil preparation and ends with crop storage and processing. Instead of bioreactors and WFI stills, plant-based biopharmaceuticals require tractors, planters and sprayers. Although the cGMPs were developed primarily for cell based production systems, many of their aspects are applicable to plant transgenic production processes, especially the need to prevent contamination and cross-contamination during production, and the need to maintain the identity of the biopharmaceutical product under production.

4.4.1. 'Pharming' Controls

Recently, the FDA issued a CMC guidance document for production of biopharmaceuticals by transgenic plants, in which they made it very clear that these bioengineered plant production processes need to be appropriately controlled (Table 36). Control over the production process is essential since gene expression in plants may be highly sensitive to environmental variables such as hydration level, nutrient level, temperature, gravity, and the intensity, direction and duration of light radiation.[97]

It needs to be stressed that GMP production activities apply to contract farms where plants are grown and to contract operations used to harvest and transport plants or plant materials.

Table 36. Production controls recommended by the FDA for biopharmaceuticals manufactured from transgenic plants[67]

Crop Growth	Establishment of specification/acceptance criteria/limits for the soil composition and potential soil contaminants that may affect the source material
	Specifications regarding the use of chemicals and limits on agricultural practices (e.g., the use of specified fertilizers, pesticides, or herbicides, and irrigation practices relative to a specified harvest time frame, etc.)
	Generate a list of expected pests that require control during the growth of the plants; and all pest-control interventions should be described in appropriate SOPs and should be documented in Batch Records
	If product expression is induced, either chemically or physically, criteria should be established to ensure that induction is performed consistently from batch to batch
	All persons involved in field growth of the product should be adequately trained to perform the duties for which they are responsible

Harvest	The harvesting process needs to be documented in production records; procedures available for determining when the harvest will occur in order to ensure lot-to-lot consistency of the source material
	Specifications for the harvested material with regard to the levels of active component, process-derived contaminants, significant endogenous impurities and adventitous agents
	Harvesting personnel need to be trained regarding plant source material quality (e.g., assessment of the disease status of plant for manual harvesting procedures); written procedures need to be established for the necessary training of personnel engaged in harvesting plants to ensure the quality of the harvested material; dedicated equipment is recommended
	Changeover procedures designed to prevent cross-contamination between harvests should be in place and documented; these procedures should include clearance of all materials and waste from the receiving area and plant material processing equipment, and cleaning/sanitization of surfaces; it is recommended that only one lot should be processed at a time
	Measures to prevent the contamination of the harvested material with equipment lubricants during processing need to be considered
Transport	The source material needs to be transported from field to initial processing site in such a way as to exclude introduction of insects, vermin, or potential surface contaminants
	Appropriate controls should be specified (e.g., stability of the product, ability to support growth of microorganisms, residual soil content, presence of foreign material, insects, vermin)
	Source material should be stored under appropriate conditions to ensure that decomposition processes do not increase the concentration of contaminants above specified levels or adversely affect the desired biopharmaceutical product
Initial Processing	Procedures used to process harvested material should be validated (e.g., harvested material may be processed to lower bioburden or viability, improve its handling characteristics, bulk consistency, and/or extractability using various procedures including washing, sanitizing, milling of grain, shredding of leaves, homogenization of plant material)

The EMEA has a similar CMC guidance document for plant transgenics which describes the process variable content that the EMEA expects to see when describing the bioengineered plant production process (Table 37).

Table 37. CMC content required by EMEA for describing a transgenic plant production process[97]

Scale, items of equipment, transport and storage.

Propagation steps. The number of generations should be clearly defined for elite, transgenic and production plant lines with reference to the documented genetic stability of the process.

Procedures for the detection and removal of undesirable plants generated by sexual propagative techniques. When self-fertilised, the progeny of plants what are heterozygous for a transgene consist of a mixture of plants which lack the transgene, plants which are heterozygous for the transgene, and plants which are homozygous for the transgene. Similarly, when a plant which is heterozygous for a transgene is crossed with an untransformed elite plant, the progeny consist of a mixture of heterozygous and untransformed individuals. As it should not be assumed that the different genotypes yield biomass which is acceptable for inclusion in crude harvests, validated techniques for the detection and removal of undesirable plant genotypes should be developed and described.

Procedures for the detection and removal of weeds and pests.

Procedures for plant health status monitoring, plus actions to be taken in case of disease.

Conditions and durations for storage of isolated in-process materials.

Cultivation variables. These are likely to include: (a) planting technique and location, taking into account environmental conditions including sensonality and nature of neighbouring flora, (b) soil type and radioactivity. (c) plant hormone and fertiliser application, (d) pesticide application, including the use of chemical and biological agents; (e) qualifications of, and training for, personnel involved in crop monitoring and cultivation.

In-process monitoring of production consistency. Parameters such as yield (which is conventionally expressed as a percentage of total extractable protein), and mRNA production (using techniques such as northern blotting or RT-PCR) should be considered for routine monitoring.

Criteria for initiation of harvesting

Harvesting technique

Procedures for the immediate manipulation of biomass once harvested, including transport and storage arrangements, and mechanical, physical, chemical and biological treatments applied.

Requirement to address the viral safety of the medicinal product. Potential sources of viral contamination during production: (1) plant virus infection of the transgenic plant line, (2) insect, bird and animal excreta, (3) insect, bird and animal carcases entrained during harvesting, (4) organic fertilizer, (5) production personnel and equipment, and (6) materials such as growth promoters introduced during production.

Any materials introduced during production should be identified and demonstrated to be in compliance with minimizing the risk of transmitting animal spongiform encephalopathy agents.

4.4.2. USDA/APHIS Protecting the Gene Pool

Containment is a major concern for bioengineered plants. The Biotechnology Regulatory Services Division (BRS) within the Animal and Plant Health Inspection Service (APHIS) of the US Department of Agriculture (USDA) oversees the importation and interstate movement of bioengineered plants and infectious plant vectors as well as the release of these entities into the environment (i.e., outside of a contained facility, such as a greenhouse, laboratory, or fermentor). The manufacturing company must receive a permit from APHIS/BRS prior to engaging in production (7 CFR 340) and must meet all of the stated environmental controls. A copy of this permit is to be included in the IND and/or BLA/NDA. APHIS/BRS has strict requirements for minimizing gene flow during crop cultivation (Table 38).

Table 38. APHIS procedures to prevent gene flow from transgenic plants[98, 99]

Transgenic	Gene Flow Controls
Barley	Barley is self-pollinating. To further minimize any gene flow via pollen, transgenic barley will be (1) planted no closer than 500 feet from other barley plants and (2) isolated temporally by sowing the seed no less than 28 days before, or 28 days after any nontransgenic seed that is sown within a zone extending from 500-1000 feet away from the transgenic plot. The transgenic plot will be monitored for 2 years following the field planting to detect and eliminate any volunteer barley plants.
Corn	The transgenic corn must be planted at sites that are at least 1 mile away from corn seed production and must ensure that any corn from previous seasons is harvested and removed in a radius of 0.25 mile of the transgenic corn plot, before the transgenic corn is sown. The land within 50 feet of transgenic plant area must remain fallow during the test. For open-pollinated transgenic corn, no other corn plants can be grown within a radius of 1.0 mile of the transgenic plants, at any time. Transgenic corn must be planted no less than 28 days before, or 28 days after the planting dates of any other corn that is growing within a zone extending from 0.5 to 1.0 mile of the transgenic corn plants.
Rice	To prevent gene flow via pollen, transgenic rice must have (1) border rows of nontransgenic rice to dilute pollen from transgenic rice plants, (2) a minimal isolation distance from other non-regulated rice of 100 feet, and (3) a temporal isolation distance of at least 14 days difference in the anticipated flowering period to the closest rice fields outside the 100 feet isolation zone.
Tobacco Mosaic Virus (TMV) (transient expression in Tobacco)	The closest commercial tobacco production site should grow TMV-resistant cultivars. A strip of fallow ground should be maintained around the field of tobacco that is to be infected with the genetic engineered virus and other weeds that are hosts for the virus should be controlled on site by either herbicide application or roguing. A non-host species should be grown in the arable land adjacent to this strip of fallow ground to act as a barrier to the spread of the virus to other fields. Inoculation of the tobacco plants with the virus should be performed by spray applicators that control the distribution of the virus. All farm implements that come in contact with infected plants should be cleaned thoroughly to ensure inactivation of any residual engineered virus. The field site must be redisked at least twice after final harvest to facilitate natural decay of plant material. The following year a non-host species should be grown in the field to allow additional time for any remaining engineered virus to biodegrade.

The seed, crops and grain need to be controlled in a manner that ensures that the transgenic plants or the biopharmaceutical product do not enter either the animal or human food chain.   Some of the controls required by the USDA and the FDA are presented in Table 39.

Table 39. List of some APHIS and FDA requirements for ensuring that plant-based biopharmaceuticals do not enter the animal or human food chain [67, 99]

Seed Stock	Maintain a careful control over the inventory and disposition of viable seeds
	Account for total yield of seed (e.g., by weight or by volume)
	Store seed stocks in aliquots of appropriate volume to allow reasonably accurate accounting of use and disposition
	Record the amount and disposition of any withdrawals from the seed bank
	Seed stocks should be prominently labelled in accordance with the permit issued by APHIS/BRS
Field-grown Plants	Must have a permit from APHIS/BRS
	Account for the seed that is transferred from seed bank storage to the filed for planting, or for archiving
	Document the size and location of all sites where the bioengineered plant will be grown, of the control of the pollen spread, and of the subsequent use of the field and destruction of volunteer plants in subsequent growing seasons
	Fields should be unambiguously identified, such as by Global Position Satellite (GPS) markers
	Consider the use of perimeter fencing
	APHIS requires that planters be dedicated to use in the permitted test site for the duration of production
	While tractors and tillage attachments, such as disks, plows, harrows, and subsoilers do not have to be dedicated, they must be cleaned in accordance with protocols approved by APHIS
	APHIS requires the use of dedicated facilities for the storage of equipment for the duration of production; facilities must be cleaned according to APHIS-approved protocols
Harvested Material	APHIS requires that harvesters be dedicated to use in the permitted test site for the duration of production
	Use transport confinement procedures from the field to the production facility per 7 CFR 340
	Label containers clearly to indicate the material inside
	Document the quantities of material leaving the growing facility and reconcile with that arriving at the processing facility

Processing Facilities	Bioengineered plant materials should not be processed in facilities that are also used for the production of food or feed, such as grain mills, without prior consultation with APHIS/BRS and FDA
Waste	Bioengineered biomass should be disposed in a manner to ensure that the material does not enter the human or animal food chain

In addition, the FDA recommends strategies that would allow the bioengineered biopharmaceutical plant line to be readily distinguished from its food or feed counterpart.[67] Such strategies include the use of genetic markers that alter the physical appearance of the plant (e.g., a novel color or leaf pattern), or change the conditions under which a plant will grow (e.g., the use of auxotrophic marker gene).

The FDA also recommends strategies that would reduce the likelihood of unintended exposure to a regulated product by restricting the expression of the biopharmaceutical to a few specific plant tissues (e.g., the use of tissue specific promoters) or by restricting the conditions under which the product will be expressed (e.g., use of an inducible promoter). For plants that outcross, consideration should be given to growing them in regions of the country where little or none of its food/feed counterparts are grown.

Europe has a similar containment concern and they require transgenic plants to meet the legislation requirements of Genetically Modified Organisms (GMOs)[67]. This legislation, Directive 2001/18/EC of the European Parliament, applies to deliberate release of GMOs into the environment. Furthermore, this directive requires the progressive elimination of antibiotic marker genes from the transgenic plants.

Until the industry can demonstrate its compliance with containment and control, use of transgenic plants to product biopharmaceuticals will undergo rigid regulatory, as well as public opinion, scrutiny.

5. WHAT CAN GO WRONG

With all of the things that need to be properly controlled during production of a biopharmaceutical, companies in their haste can fail to meet the requirements expected by the regulatory agencies. This usually results in inspectional issues when the FDA arrives at the manufacturing facility. Table 40 presents some production process control deficiencies recorded by the FDA.

Table 40. Some production process deficiencies recorded by FDA inspectors[81, 100, 101]

Establishment Inspection Report, IDEC Pharmacetuicals, May 19-June 3, 1997

During the manufacturing campaign, five contamination events were due to loss of integrity in the bioreactors, overhead drive spinner flask (ODSF), or transfer line; corrective action did not include routine integrity testing of the transfer line, vent and sparge filters, ODSFs, and the bioreactors after the manufacturing run

Regarding the sterilization/sanitization of the bioreactor: (1) thermocouple temperatures were not correlated with the resistance temperature device data, (2) pressure testing of the bioreactor is inconsistent in that post-CIP bioreactors are tested at 20-21 psi and post-SIP pressure testing is at 3 psi, (3) temperature and pressure are not monitored or recorded during the sanitization/sterilization of transfer lines, and (4) the Tempilstik used to monitor temperatures in valves has not been qualified

Not all clean steam lines supplying bioreactors and jacketed tanks in cell culture are sloped to drain

Switches controlling the bioreactor agitator are not protected and can be switched off accidentally

Establishment Inspection Report, IDEC Pharmaceuticals, October 6-8, 1997

Bioreactor operating parameter ranges for pH and dissolved oxygen are not consistent with retrospective validation data

During the campaign on 8 separate occasions the parameters of the cell culture were not recorded within the specified times even though there is a window within which these readings must be taken

The firm has not established written procedures pertinent to periodic assessment for the need for repassivation of stainless steel equipment (e.g., bioreactors, portable holding tanks, spray. balls/wands and stainless steel product transfer lines)

FDA483, Genentech, Inc., February 7-14, 2000

Hoses used for transferring in-process product form the fermenter to the filtration unit are cleaned in place after use and stored in the lower fermentation room, open to the environment (ends not covered). The room is monitored as a controlled unclassified environment; the hoses are not cleaned again or sterilized prior to use

Not all vent filters used in fermentation and purification are integrity tested post-use; vendor integrity testing is confirmed by review of the certificate of analysis

Controlled, non-classified areas are assigned alert limits for viable and non-viable particulate monitoring but are not assigned action limits; these areas include Recovery, fermentation, Inocula/Fermentation, Media and Buffer Preparation, Sterile Equipment Storage, and Supply Corridors; operations in recovery and fermentation areas are not entirely closed

There was no formal documentation demonstrating the compatability of the validation of the effectiveness of disinfectants between the South San Francisco Facility and the Vacaville Facility

6. STRATEGIC CMC TIPS FOR PRODUCTION

It has been said that the regulatory expectations at each clinical development stage often exceed the best efforts of those charged with translating senior management directives into direct actions. Therefore, careful planning with a focus on the 'must haves' is critical to manage the CMC minimum continuum for the production process of a biopharmaceutical.

The production process for a biopharmaceutical at an early clinical stage will be basic and somewhat unpolished. The focus needs to be on ensuring safety for the expressed product, as indicated by Figure 10.

Figure 10. Minimum CMC continuum for a biopharmaceutical production process

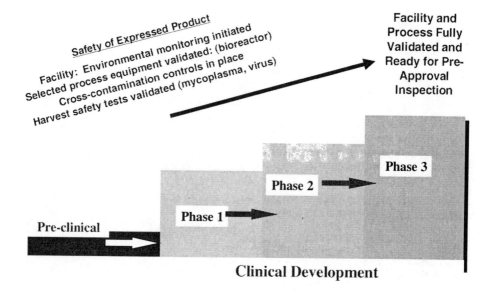

While the list in the Figure will vary with the type of biopharmaceutical being produced, the recombinant host chosen, the complexity of the facility operations, and even the person drawing up the list, each of the activities selected should be related to ensuring that the expressed biopharmaceutical is safe for the patient. By time of the BLA/NDA filing, all CMC regulatory compliance issues related to the production process are to be completed.

Purification of the Biopharmaceutical

"The extent of purification of recombinant DNA products should be consistent with the intended use of the product. Drugs and biologics which are to be administered repeatedly or at high concentrations should be adequately pure to prevent the development of undesired immune or toxic reactions to contaminants. The purification process should be designed to specifically eliminate detectable viruses, microbial and nucleic acid contamination and undesirable antigenic materials."

1985 FDA guidance on the production and testing
of recombinant DNA-derived products[2]

1. GOALS: PURITY, RECOVERY AND CONSISTENCY

From the harvested production, the expressed biopharmaceutical needs to be purified in sufficient quantity to produce the bulk active pharmaceutical ingredient (API). Regardless of the type of production system used or its scale, the products will be purified using procedures common to either protein or DNA molecules.

The goal of the purification process is to effectively separate the product from both its impurities and any putative adventitious agent contamination, without unacceptably altering the product.

Each harvested production system presents similar but unique challenges for the purification process; for example, if recombinant mammalian or human cells are used, the purification process must be robust in its ability to remove viral contaminants. It should be noted that for some gene therapy products (e.g., a patient's cells which have been genetically altered with a viral vector), purification may not even be possible.

1.1. Two Major CMC Regulatory Concerns for Purification

There are two major CMC regulatory concerns about the purification processes used for biopharmaceuticals:

1. Proper Design

    The purification processes must be adequate designed to sufficiently separate the biopharmaceutical product from the many components present after the harvest of the production process, including the capacity to remove putative viruses that might accidentally contaminate the production system.

2. Consistent Performance

    The purification processes must have adequate procedures and controls in place to ensure that the process is consistently operated, and that it yields a consistent product of the desired quality.

It is the responsibility of each manufacturer to demonstrate to the regulatory agencies that their specific purification process is adequately designed and controlled, yielding the desired product.

Note, the regulatory agencies, in contrast to the manufacturer, are not that concerned about the level of product yield, except when that yield might impact the ability of the manufacturer to produce enough product for market launch.

1.2. Need for High Recovery of a Pure Product

A manufacturer does not really have a useable purification process until it can be shown that the process yields a suitably high recovery (many purification processes provide overall recoveries of the biopharmaceutical in the 60% level or higher) and acceptable product purity (many purification processes are targeted at the 95% purity level or higher), every time the process is run.

Most purification processes at Phase 1 will not meet these criteria, and they are not required to do so. Unfortunately, many biopharmaceutical companies rush into the clinic with a purification process, fail to commit to do process development alongside their clinical development; and if they find clinical value, they are then faced with a purification process that either may not meet regulatory expectations or cannot be run economically. Continual development and improvement of the purification process during the clinical drug development stages is essential for eventual commercial success.

1.3. What is a 'Purification Process'?

The goal is to go from the harvested production material to an isolated, pure biopharmaceutical product. Ideally, this should occur under conditions that are stable, and conditions that produce the desired pure biopharmaceutical in the required amounts over a realistic timeline. And furthermore, purify the biopharmaceutical under conditions that yield a safe product and one with appropriate quality characteristics.

The purification process for a biopharmaceutical, which starts with the harvested production material, may or may not first require extraction, may or may not require physical separation, but will always require one or more chromatographic process steps, to eventually yield the purified API, as illustrated in Figure 11.

Figure 11. Steps of a biopharmaceutical purification process

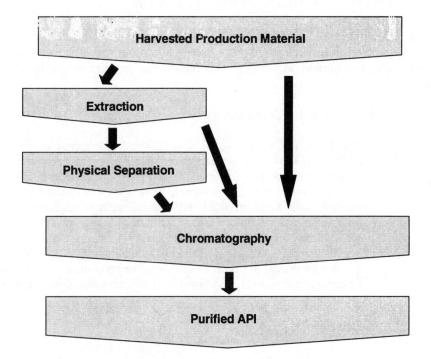

For harvested biopharmaceuticals present in aqueous solution (e.g., recombinant proteins and monoclonal antibodies expressed in either cell-based bioreactor production systems or milk from transgenic animals), purification can readily proceed. However, for harvested biopharmaceuticals not present in aqueous solution (e.g., seeds from transgenic plants), further processing (such as milling the seeds into a fine flour) occurs first, followed by aqueous extraction of the biopharmaceutical, which is then ready for purification to proceed.

The purification of the product from its impurities involves multiple process steps incorporating various types of physical separations and chromatographic methods. For biopharmaceuticals harvested from production systems that can amplify adventitious virus, the purification process is also designed to contain viral inactivation (e.g., low pH treatment) and viral removal steps (e.g., nanofiltration) in order to provide adequate clearance of putative viral contaminants from the production process.

After the purification process, the bulk active pharmaceutical ingredient (API) is typically placed in a suitable container and stored until needed for later formulation and drug product manufacturing.

1.3.1. Physical Separations Methods for Biopharmaceuticals

One of the primary physical separation methods for recombinant proteins and monoclonal antibodies is precipitation, with either the biopharmaceutical or the protein impurities being precipitated, followed by centrifugation to separate the precipitate from the solution:

- To Precipitate the Biopharmaceutical

    The properties of the solution are altered, for example, either by addition of salts such as ammonium sulfate or by addition of organic polymers such as polyethylene glycol (PEG). By increasing either the salt concentration or the non-ionic polymer PEG concentration, the additives are excluded from the protein surfaces resulting in their preferential hydration, thus facilitating hydrophobic surface associations between protein molecules. First protein aggregation begins and eventually protein precipitation. After the biopharmaceutical has been precipitated, the precipitate is collected by centrifugation, washed, and then the biopharmaceutical is resuspended in solution.

- To Precipitate the Impurities,

    The properties of the solution are altered, for example, by addition of organic complexants. By adding the short chain fatty acid, octanoic acid, also called caprylic acid, to the protein mixture at low pH, acidic protein impurities increase in hydrophobicity, resulting in their precipitation. By adding the organic base, ethacridine, to the protein mixture at high pH, basic protein impurities form stable insoluble ionic complexes with the chemical, resulting in their precipitation. After the impurities have been precipitated, the precipitate is collected by centrifugation and discarded. The solution containing the biopharmaceutical is kept.

A second physical separation method for biopharmaceuticals is filtration, in which a fixed membrane pore size is used to allow molecules of a smaller size to flow through the membrane while retaining the larger molecules or organisms. Depending upon the pore size, the biopharmaceutical could either flow through the membrane (e.g., when performing sterile filtration) or be retained by the membrane (e.g., when performing a buffer exchange).

1.3.2. Chromatographic Purification Methods for Biopharmaceuticals

There are several chromatographic separation methods available for recombinant proteins and monoclonal antibodies:

- Size Exclusion Chromatography (SFC)

   SEC separates proteins by molecular size, by taking advantage of column resins having a specified pore diameter. Proteins too large in size to enter the resin pores are excluded and move through the column rapidly, eluting at the beginning of the chromatogram, the void peak. Proteins having access to the pores are said to be included. Their separation characteristics are determined by the pore size distribution within the pore volume. Proteins at the upper molecular size end are included only in the larger pores, which gives them a shorter path length through the column and earlier elution. Proteins at the lower molecular size end interact with all of the pores, which give them a longer path length through the column and later elution.

- Ion Exchange Chromatography (IEC)

   IEC separates proteins by their charged residues. Charged residues on protein surfaces include the side groups of amino acids, the a-amino and a-carboxyl termini of the chains, and sialic acid residues on glycoproteins. These residues are amphoteric, making the sign and net charge on proteins a function of pH. The pH at which a protein's positive charge balances its negative charge is its isoelectric point (pI). A protein binds to an anion exchanger (resin bearing negative charge) at pH values above its pI. A protein binds to a cation exchanger (resin bearing positive charge) at pH values below its pI. The bound proteins are eluted either by altering the pH of the column buffer, by addition of competing ions, or both.

Membrane-based chromatography is a form of IEC. Rather than a column packed with resin beads, polymeric ion exchange membranes contained in cassettes are used. Cation exchange membranes capture proteins while allowing DNA and endotoxins to flow through; anion exchange membranes retain DNA and endotoxins while allowing the protein to flow through. These membranes are single-use and disposable.

- Hydrophobic Interaction Chromatography (HIC)

   HIC utilizes two fundamental principles. The first principle is that salts are preferentially excluded from both protein surfaces and chromatography resins. Thus with increasing salt concentration in the column buffer, it becomes favourable for proteins to bind to the resin. The second principle is that hydrophobic interactions are a strong attractive force in salt solutions. Therefore, strongly hydrophobic proteins bind more strongly to the resin than weakly hydrophobic proteins. By removing the salt from the column buffer, the bound proteins elute according to their hydrophobicity, from lower to higher. Sometimes the addition of organic solvents to the column buffer is necessary to elute the bound strongly hydrophobic proteins.

- Affinity Chromatography (AC)

   AC exploits the interaction of bound ligands on a chromatographic resin with proteins in solution. The ligands can be chemical or biochemical. Chemical ligands include chelated metals (e.g., nickel to bind to histidine sites on the protein) and dyes (e.g., Cibachron blue, also known as blue-2, to bind to strongly basic proteins). The most widely used biochemical ligand is Protein A (a cell wall component of *Staphylococcus aureas* that binds strongly to IgG monoclonal antibodies). The protein solution is passed through the affinity column, the specific proteins bind to the column resin, the column resins are washed, and then the bound proteins are eluted.

- Reversed Phase Chromatography (RPC)

   RPC also separates proteins by their hydrophobicity. The column resin is coated to provide a non-polar hydrophobic surface. In the presence of organic solvents, it becomes favourable for the proteins to bind to the resin. By increasing the concentration of the organic solvents and decreasing the concentration of water in the column buffer, the bound proteins elute according to their hydrophobicity, from lower to higher.

The design of the purification process will include some combination of these chromatographic steps which will vary with the biopharmaceutical product needing to be purified.

For therapeutic IgG monoclonal antibodies, where many have already been approved by the FDA for the market and many more are in clinical development, it is now possible to consider a generic purification process for these biopharmaceuticals:

[Cell Culture Harvest]

↓

Protein A Affinity Chromatography

↓

Viral Inactivation

↓

Polishing Chromatographic Step (IEC or HIC)

↓

Viral Filtration

↓

API

For the DNA-based biopharmaceuticals, the choice of chromatographic methods is currently limited. Anion exchange chromatography, which is widely used to separate proteins from residual DNA, finds the reverse application for the DNA-based vectors, in which residual protein is desired to be separated from DNA product. Further purification of the DNA can be accomplished by using one or more of the other chromatographic processes (such as SEC, HIC or RPC).

1.4. Purification Processes Familiar to the FDA

Regulatory agencies, since they review and approve the purification processes to grant market approval, are very familiar with purification processes involving the different types of physical separation and chromatographic methods. Table 41 presents some specifics for purification processes for biopharmaceuticals.

Table 41. Purification process information from FDA approved biopharmaceuticals[63, 71, 102]

Palivizumab (Synagis®), Humanized Monoclonal Antibody

Harvested Cell-Free Conditioned Medium
⇩
Cation Exchange Chromatography
A product capture step optimized for volume, and reduction in bovine serum albumin (BSA) and DNA
⇩
Benzonase Treatment
Enzymatic treatment to reduce the level of residual DNA
⇩
Protein A Affinity Chromatography
Removal of Benzonase in addition to other process and cell-derived contaminants
⇩
Nanofiltration
Viral removal process step.
⇩
Anion Exchange Chromatography
Further reduction in process contaminants such as bovine IgG, transferrin and endotoxin.

(All purification operations are carried out in environmental rooms each with an air classification of 10,000 and terminal HEPA filtration. Process operations are also segregated according to cleanliness level, and pre- and post-viral clearance procedures. Full gowning is required for personnel in the purification suite and consists of a sterile clean room gown, boots or shoe covers, hair cover, gloves, and face mask.)

Infliximab (Remicade®), Chimeric Mouse/Human Monoclonal Antibody

Harvested Cell-Free Conditioned Medium
⇩
Protein A Affinity Chromatography
A product capture step optimized for volume and removal of serum albumin (BSA)
⇩
Solvent/Detergent Treatment
Virus inactivation process step
⇩
Cation Exchange Chromatography
⇩
Nanofiltration
Virus removal inactivation step
⇩
Anion Exchange Chromatography
⇩
Anion Exchange Chromatography
Further reduction in bovine IgG impurity level

<u>Aldesleukin (Proleukin®), Recombinant Human IL-2</u>

Harvested Particulate Mattter
⇩

SDS Buffer Treatment
To extract the protein into aqueous solution
⇩
Protein Reduction
⇩
Size Exclusion Chromatography
⇩
Protein Oxidation
To refold the protein into the correct configuration
⇩
Reversed Phase High Performance Liquid Chromatography (RP-HPLC)
⇩
Precipitation
⇩
SDS Buffer Treatment
To solubilize the protein product
⇩
Size Exclusion Chromatography
⇩
Diafiltration

2. ADEQUATE DESCRIPTION OF THE PURIFICATION PROCESS

As indicated previously in the chapter on production, the FDA expects the manufacturer to make the following statement in their Phase 1 IND submission: "the sponsor believes that the manufacturing of either the drug substance or the drug product does not present any signal of potential human risk."[16] To support such a statement, the manufacturer needs to describe the purification process for the biopharmaceutical. This description needs to be accurate and of sufficient depth to allow the FDA reviewer to conclude that the purification process is indeed safe, does what it has been designed to do, and does not impose any safety risks to the purified biopharmaceutical.

The amount of information about the purification process that the regulatory agencies expect to see included in the regulatory submissions increases as the clinical development advances. As stated in the FDA regulations, 21 CFR 312.23(a)(7)(i):

> "Although in each phase of the investigation sufficient information is required to be submitted to assure the proper identification, quality, purity and strength of the investigational drug, the amount of information needed to make that assurance will vary with the phase of the investigation, the dosage form, and the amount of information otherwise available."[34]

Keep in mind, that for biopharmaceuticals, the regulatory expectation is that "more information may be needed to assess the safety of biotechnology-derived drugs" compared to chemically-synthesized drugs.[16]

2.1. During Clinical Development

The minimum CMC continuum applies to purification processes. At each stage of clinical development, the regulatory agencies expect that the minimum CMC requirements will be met, and that these requirements will increase and tighten as clinical development progresses.

2.1.1. Phase 1 IND Submission

For the Phase 1 IND, the amount of information needed to describe the purification process is expected to be 'brief':

"The full address of the manufacturer of the clinical trial drug substance should be submitted.

A brief description of the manufacturing process, including a list of the reagents, solvents, and catalysts used, should be submitted. A detailed flow diagram is suggested as the usual, most effective, presentation of this information."[16]

But what is 'brief'? My recommendation is that sufficient information about the purification process needs to be provided to assess safety of the biopharmaceutical. Ask yourself, what would you need to know about the purification process to make this assessment? That's what should be provided. Clearly, this doesn't mean submitting the batch production records. But it does mean summarizing the specifics of those batch production records and including them in the submission. An example of the level of CMC detail that could be provided in the submission at this stage of clinical development is presented in Table 42. This table only presents one step of the entire purification process, so similar sections would need to be written for each of the purification steps. In addition, sections would need to be written for the acceptance criteria for all of the raw materials used in purification, for any specialized preparation of purification buffers, for the viral safety steps included in the purification process, and for the in-process controls to be used. In addition, the requested detailed flow diagram needs to be included.

Table 42. Illustration of the level of CMC content for a Phase 1 IND submission: One chromatographic process step

Concentration and Diafiltration

Adjust pH to 7.0 using 1 M sodium phosphate dibasic or 1 M hydrochloric acid. Concentrate the pooled fractions to 1-2 mg/mL using hollow fiber cartridges with a 10,000 MW cutoff. Diafilter the concentrate against 6 volumes of 20 mM sodium phosphate buffer, pH 7.0.

Phenyl Sepharose Chromatography

Equilibrate Phenyl Sepharose column (20 x 100 cm, bed height of 80 cm) with aqueous buffer containing 1.2 M ammonium sulfate and 10 mM sodium phosphate pH 7.0. Load the diafiltered product (50-200 g) onto the column. The product binds to the column resin. Elute the column using a gradient of 1.2 M to 0.0 M ammonium sulfate in 10 mM sodium phosphate. The column effluent is monitored by optical absorbance at 280 nm and elution fractions are collected. Pool the fractions using SDS-PAGE analysis of the collected fractions. Filter the pooled fractions through a 0.2 micron capsule filter. Hold pool at 2-8°C for up to 12 hours. The column is regenerated by treatment with the following successive solutions: 2 M NaCl, water, 30% isopropyl alcohol, water, and 1 N NaOH. The column is stored in either 0.1 N NaOH for less than one week or 10 mM NaOH for greater than one week.

2.1.2. Phase 2 IND Submission

It is expected that additional knowledge about the purification process, especially how to more adequately control it and to make it more robust, will be obtained during the clinical development stages. The FDA expects this additional information to be included in a Phase 2 submission, and also updates of any purification process changes:

> "The general description of the synthetic and manufacturing process (e.g., ... purification) described in Phase 1 should be updated from a safety perspective if changes or modifications have been introduced.
>
> An updated detailed flow diagram for the synthesis or manufacturing process should be provided.
>
> Reagents (including solvents and catalysts), equipment (... columns), and provisions for monitoring and controlling conditions used in each step should be provided.
>
> To the extent possible in Phase 2, sponsors should document that the manufacturing process is controlled at predetermined points and yields a product meeting tentative acceptance criteria. Although in-process controls may still be in development, information on in-process controls for monitoring adventitious agents should be provided."[36]

2.1.3. Phase 3 IND Submission

By the time of the Phase 3 submissions, a much greater amount of purification description is expected to be filed:

"Updates of the acceptance criteria and analytical procedures for assessing the quality of starting materials should be provided. A table listing all reagents, solvents, catalysts should be submitted that includes (1) the grade of each material used, (2) the specific identity test performed, (3) the minimum acceptable purity level, and (4) the step of the synthesis and manufacturing process in which it is used. For special reagents (e.g., sera, enzymes, or proteins of animal origin), a more comprehensive list of tests, screening and acceptance criteria may be needed. In critical cases (e.g., monoclonal antibodies configured in affinity matrices), a full description of the manufacturing process may also be needed.

An updated detailed flow diagram should be provided. Reagents, equipment (e.g., ... columns) and monitored and controlled conditions used in each step should be identified.

A general step-by-step description of the synthesis and manufacturing processes should be provided. Relevant information should indicate batch size (range), the type of reaction vessel, relative ratios of reactants, general operating conditions (time, temperature), in-process controls (complete description of analytical procedures) and literature references for novel reactions or complex mechanisms.

Controls at selected stages in the synthesis and manufacturing process that ensure reaction completion, identity, and purity or proper cell growth should be described. The acceptance criteria and analytical procedures should be described for isolated intermediates that require control. Tentative acceptance criteria can be used to allow for flexibility in the development process, but should fulfill the primary purpose of quality control. The description of the analytical procedures should be brief, and appropriate validation information should be available on request. Reprocessing procedures and pertinent controls should be described."[36]

2.2. Preparing the BLA/NDA Submission

The 'complete' description of the purification process is required in the market approval submission. And when the FDA says 'complete', they mean it. In fact, if this purification information is not complete, it can be a cause for the FDA to refusal to file the submission.[31] Some of the CMC information on the purification process that is required is presented in Table 43.

Table 43. CMC information to be included in the BLA/NDA dossier to describe the purification process for recombinant proteins and monoclonal antibodies[40]

Identification of Manufacturer(s)	The application should include the name(s), address(es), FDA registration number, and other pertinent organizational information for each manufacturer performing any portion of the manufacture of the drug substance. A brief description of the operations performed at each location, the responsibilities conferred upon each party by the applicant and a description of how the applicant will ensure that each party fulfills their responsibilities should be submitted.
Floor Diagram(s)	For each manufacturing location, a floor diagram should be included that indicates the general facility(ies) layout. This diagram need not be a detailed engineering schematic or blueprint, but rather a simple drawing that depicts the relationship of the subject manufacturing areas, suites, or rooms to one another, and should indicate other uses made of adjacent areas that are not subject of the application. This diagram should be sufficiently clear such that the reviewer may visualize the flow of the production and would be able to identify areas or room 'proximities' that may be of concern for particular operations, e.g. segregation of pre and post viral inactivation material and operations, segregation of animal facilities, etc. Room numbers or other unique identifiers should be provided, however it is not necessary to include the location of processing equipment with rooms and areas.
Other Products	A comprehensive list of all additional products to be manufactured or manipulated in the areas used for the product should be provided. The applicant should indicate in which rooms the additional products will be introduced and the manufacturing steps that will take place in the room. An explanation should be given as to whether these additional products will be introduced on a campaign basis or concurrently during production of the product which is the subject of the application. Any additional products that may share product contact equipment with the product in question should be indicated (dedicated vs multi-use equipment should be delineated). A brief description should be provided as the type and developmental status of the additional products.
Contamination Precautions	For all areas in which operations for the preparation of cell banks and product manufacturing are performed, the following information concerning precautions taken to prevent contamination or cross-contamination should be provided: • Air quality classification of room or area in which operation is performed, as validated and measured during operations • A brief, narrative description of the procedures and/or facility design features for the control of contamination, cross- contamination and containment (air pressure cascades, segregation of operations and product, etc.) – this is of particular importance for multi-use areas • General equipment design description, e.g., does design represent and open or closed system or provide for a sterile or non-sterile operation • A description of the in-process controls performed to prevent or to identify contamination or cross-contamination
Raw Materials and Reagents	A list of all components used in the manufacture of the drug substance, and their tests and specifications or reference to official compendia should be provided. For purchased raw materials representative certificates of analysis from the supplier(s) and/or manufacturer's acceptance criteria should be included in the submission. Process gases and water are considered raw materials. A list with tests and specifications of all special reagents and materials used in the manufacture of the drug substance, e.g., buffers, monoclonal antibodies, preservatives, etc. should be submitted. In cases where an ancillary biological product is used in the

manufacture of the drug substance (e.g., a monoclonal antibody used in affinity chromatography), a detailed description of the preparation and characterization of the reagent should be submitted.

A description of the tests and specifications for materials of human or animal source that may potentially be contaminated with adventitious agents, e.g., mycoplasma, BSE agent, and other adventitious agents of human and animal origin should be submitted. Validation data or certification supporting the freedom of reagents from adventitious agents should be included in the submission.

Process Flow Charts	A complete visual representation of the manufacturing process flow should be provided. This flow chart should indicate the step in production, the equipment and materials used, the room or area where the operation is performed and a complete list of the in-process controls and tests performed on the product at each step. This diagram should also include information (or be accompanied by a descriptive narrative) on the methods used to transfer the product between steps, i.e., sterile, steam-in-place (SIP) connection, sanitary connection, open transfers under laminar flow units, etc. Such transfers should be described for movement of product between equipment, areas/rooms, buildings and sites. Manufacturing steps which are computer controlled should be identified. Manufacturing steps which are computer controlled should be identified.
Purification and Downstream Processing	A detailed description of the purification and downstream processing, including a rationale for the chosen methods, and the precautions taken to assure containment and prevention of contamination or cross-contamination should be provided. In-process bioburden and endotoxin limits should be specified where appropriate. Any reprocessing using a validated reprocessing method and the conditions for batch eligibility should be described. Indication should be made as to the multi-use nature of areas and equipment (e.g., campaigning vs. concurrent manufacture; dedicated vs. shared equipment) used for these procedures. A brief description of the controls employed to ensure segregation and prevent cross-contamination should be provided.
Batch Records	A completed (executed) representative batch record of the purification process should be submitted.
In-Process Controls	A description of the methods used for in-process controls, e.g., those involved in downstream processing, should be provided.
Process Validation Studies	A description and documentation of the validation of the purification process to demonstrate adequate removal of extraneous substances such as chemicals used for purification, column contaminants, endotoxin, antibiotics, residual host proteins, DNA, and viruses, should be provided.

3. FACILITY AND UTILITY CONCERNS

The design and operation of the purification facility is the first step in understanding what will be necessary for eventual validation of the purification process.

3.1. Design and Operation

The facility could be for single-product manufacturing or for multi-product campaign manufacturing or for multi-product concurrent manufacturing. Emphasis should be placed on operational features that minimize the risk of contamination from the environment or cross-contamination from other products.

As the complexity of the activities in the production facility increase from single-product to multi-product, the regulatory agencies have greater concern about contamination and cross-contamination issues. And they will have great interest in the manufacturer's capability of controlling this through (1) segregation of operations, (2) procedural controls, (3) sharing of process equipment, and (4) air handler design.

The FDA has made it crystal clear that the biopharmaceutical manufacturing facility, and utilities must be completely validated in order to pass the pre-approval inspection for market approval. This applies equally to small-scale pilot facilities as well as to large-scale multi-product commercial facilities.[79]

All purification steps should be performed in environmentally controlled areas, and each purification step should be evaluated individually for appropriate environmental control (e.g., monitoring frequencies, appropriate alert and action limits, surface monitoring, personnel gowning, and corrective action).

The goal is to carry out the purification process without introducing adventitious contamination. According to ICH Q7A Section 18 Specific Guidance for APIs Manufactured by Cell Culture/Fermentation, the following general principles are outlined:

> "If open systems are used, purification should be performed under environmental conditions appropriate for the preservation of product quality.
>
> Appropriate precautions should be taken to prevent potential viral contamination from previral to postviral removal/inactivation steps. Therefore, open processing steps should be performed in areas that are separate from other processing activities and have separate air handling units.
>
> Appropriate controls should be taken to prevent potential virus carry-over (e.g., through equipment or environment)."[22]

Water for Injection (WFI) is used in the purification process. Because the water generation systems in biopharmaceutical facilities are complex (consisting of water treatment units, distillation units, hot and cold distribution loops and terminal filters at use points), continual testing of water used in the production process per the United States Pharmacopeia (USP) standard is required.[86]

Controls and procedures must be in place, even at the start of Phase 1 product manufacturing, to ensure that the facility and equipment are handled in a manner that minimizes the risk of microbial contamination for the biopharmaceutical product.

3.2. Environmental Monitoring

For purification processes, the FDA has not clearly defined their expectations for environmental monitoring in a guidance document. However, in a published paper, FDA reviewers elaborated on their expectations for environmental conditions for the purification areas of biopharmaceuticals:

> "Closed Purification Process Steps: Purification or other processing steps that have minimal or no potential for environmental contract should be validated as closed systems. When a purification system is shown to be closed to environmental factors, operations performed in environmentally controlled areas of at least Class 100,000 are acceptable. When closed systems are validated, alternatives to these recommendations may be considered (it may be unnecessary to maintain Class 100,000 requirements). Air in purification areas should be monitored for viable and nonviable particulate quality during dynamic conditions. Such monitoring should occur during various stages of purification for each batch because of the variability of the environmental quality introduced by personnel practices. However, in some instances a case may be made for less frequent viable and nonviable particulate monitoring. Such circumstances should be evaluated case by case, and the overall monitoring program should be designed to incorporate some degree of periodic monitoring during production operations, involving a combination of viable and nonviable contaminant monitoring methods.
>
> Open Purification Process Steps: Purification or other processing steps that expose product to environmental conditions are considered open systems and should be performed in controlled environments, meeting at least Class 100,000 conditions. As stated above, controlled environments should be monitored for viable and nonviable particulates at least weekly during dynamic conditions. However, where possible, it is recommended that air be monitored for viable and nonviable particulate quality during each processing step. When a greater potential for product contamination exists, more stringent limits and/or controls should be established. An assessment of appropriate conditions should be based on the nature of the activities being performed, presence of additional downstream processing, and risk of contamination to product. If it is determined that aseptic conditions (Class 100) are appropriate for open steps (for example, in a laminar flow hood or area), then the aseptic environment should be located within rooms that are environmentally controlled to a sufficient degree to ensure maintenance of aseptic conditions within the Class 100 environment."[91]

For clarification, clean room classifications are based on limits for nonviable and viable airborne particulate monitoring, microbial surface monitoring, and personnel monitoring; Class 100 is also referred to as Grade A or Class M3.5 or ISO14644-1 5; Class 100,000 is also referred to as Grade D or Class M6.5 or ISO14644-1 8.[92, 93] From my experience, most biopharmaceutical manufacturers design their purification areas to meet Class 10,000 (Class C or M5.5 or ISO14644-1 7) operational conditions.

These clean room classifications need to be in place even for manufacture of Phase 1 material, since the FDA will want to review this information in the IND submission. However, assignment of the action and alert limits in the environmental monitoring program can be loose. As experience is gained with the manufacturing facility and the purification process, the limits can be appropriately tightened using this historical data. The final alert and action limits will need to be set, and justified, for the BLA/NDA submission.

4. PURIFICATION PROCESS VALIDATION

Validation of the purification process operates under the same general principles as validation of the production process, discussed in the previous chapter. Validation of the purification process is absolutely necessary in order to obtain and maintain approval for marketing a biopharmaceutical product. The absence of process validation information in a filed BLA or NDA is sufficient reason for the FDA to issue a refusal to file (RTF):

> "The following are examples of basis for RTF due to insufficient description of the manufacturing processes: ... incomplete description of production of bulk, including description of fermentation, harvest and purification procedures, process validation ...."[31]

But even if it was not a regulatory requirement, it would be important for a manufacturer to perform process validation in order to minimize process failures and potentially endanger patient safety.

4.1. When Should Purification Validation Occur?

Purification validation should occur when the purification process is defined and locked in. This clearly won't occur at Phase 1 or Phase 2. Most companies define and lock in their purification process for the manufacture of Phase 3 clinical materials. Other companies wait until after the Phase 3 manufacture to do this.

.At the early stages of clinical development (i.e., Phase 1 and 2), the regulatory agencies do not expect the purification process to be validated:

> "Process validation for the production of APIs for use in clinical trials is normally inappropriate, where a single API batch is produced or where process changes during API development make batch replication difficult or inexact. The combination of controls, calibration, and, where appropriate, equipment qualification ensures API quality during this development phase."[22]

During clinical development is the time that the manufacturing process needs to be flexible and constantly being improved upon. However, it should be noted that because of the absence of a validated process during clinical development, there is much greater pressure placed on the Quality systems within a company:

> "The production of investigational medicinal products involves added complexity in comparison to marketed products by virtue of the lack of fixed routines, variety of clinical trial designs, consequent packaging designs, the need, often, for randomisation and blinding and increased risk of product cross-contamination and mix up. Furthermore, there may be incomplete knowledge of the potency and toxicity of the product and a lack of full process validation .... This increased complexity in manufacturing operations requires a highly effective Quality system."[49]

It is expected that during clinical development, the manufacturer is gaining sufficient information about the behaviour and physicochemical properties of the biopharmaceutical product, as well as the purification process itself, to begin to clearly define the critical steps of the process and to implement appropriate in-process controls. Once the purification process is locked in, typically either prior to Phase 3 manufacture or during Phase 3 clinical studies, it is these critical purification steps and in-process controls that are then validated.

Validation studies can be performed on scaled-down processes. However, it is important that production validation be confirmed on the commercial size batches to ensure that there were no significant differences with the scale-down studies.[3] For example, when scaling-up, longer processing times can result which exposes the product to conditions of buffer and temperature for longer periods.

4.2. Process Validation Concerns for a Chromatographic Step

The purification of recombinant proteins and monoclonal antibodies includes several chromatographic column-based purification steps. These steps are based on different properties of the biopharmaceutical, such as size, charge, hydrophobicity and affinity. These operations are carried out in a packed column connected to a pump to move fluids through the column and a detector to monitor the effluent stream of the column.

Before chromatographic validation can proceed, the following components must be addressed:

- Quality and consistency of chromatographic resin

    Acceptance criteria for the incoming resin including identity, safety and functionality tests

    Shelf-life of resin

- Quality and consistency of column buffers and solutions

Acceptance criteria including identity and safety of all chemical reagents (including water quality)

Bioburden and endotoxin controls

- Equipment Qualification

  IQ and OQ

  SOPs covering each column operation – packing, equilibrating, loading, cleaning, sanitizing storage

- Assay Validation

  Test methods to be used for the validation are themselves validated

Validation of the column performance can then begin. The chromatographic parameters need to be set and then validated for:

- The intended purpose for the chromatography step (e.g., removal of specific impurities)

- Absence of impact on the product (e.g., not de-natured or aggregated; acceptable recovery)

- Leachables from the resin into the product stream (especially after storage and upon reuse)

- Cleaning of the packed column (e.g., absence of product and impurity carryover; control of contaminants such as endotoxin, microbes and virus)

The FDA offers the following guidance on what is required to validate a chromatographic process step:

- Column Cleaning

  "For columns used in purification steps, ensure the firm has procedures in place for cleaning and storage of the column housings and column media and that the procedures are followed."[103]

  "Review SOPs for equipment cleaning procedures and determine if they are validated and followed. The validation should have included an evaluation of the cleaning efficacy of dedicated equipment and assurance that detergents, if any, are removed during cleaning. The firm should have established residual limits and be using an assay appropriate to detect any contaminants. For multi-use equipment, the cleaning validation should have established that there is no carry-over between products such that the safety and efficacy of subsequent products is compromised. There should be periodic monitoring of the validated cleaning procedure to ensure continued cleaning effectiveness." [103]

- Column Reuse

  "Often columns are regenerated to allow repeated use. Proper validation procedures should be performed and the process should be periodically monitored for chemical and microbial contamination." [3]

  "There should be established life span for each column type (i.e., number of runs or months of use). Ensure that columns are not used beyond their established life span." [103]

  "Limits should be prospectively set on the number of times a purification component (e.g., a chromatography column) can be reused. Such limits should be based upon actual data obtained by monitoring the components performance over time." [29]

- Purpose for Chromatographic Step

  "Demonstration of the ability of the purification scheme to remove adventitious agents and other contaminants, by means of a clearance study." [29]

  "Manufacturers should have validation reports for the various key process steps. For example, if an ion-exchange column is used to remove endotoxins, there should be data documenting that this process is consistently effective. By determining endotoxin levels before and after processing, a manufacturer should be able to demonstrate the validity of this process. It is important to monitor the process before, during, and after to determine the efficiency of each key purification step. "Spiking" the preparation with a known amount of a contaminant to demonstrate its removal, may be a useful method to validate the procedure." [3]

- Small-Scale Studies

  "Typically, manufacturers develop purification processes on a small scale and determine the effectiveness of the particular processing step. When scale-up is performed, allowances must be made for several differences when compared with the laboratory-scale operation. Longer processing times can affect product quality adversely, since the product is exposed to conditions of buffer and temperature for longer periods. Product stability, under purification conditions, must be carefully defined. Manufacturers should define the limitations and effectiveness of the particular step. Process validation on the production size batch will then compare the effect of scale-up. Manufacturers may sometimes use development data on the small scale for validation."[3]

- Production-Scale Requirement

    "However, it is important that validation be performed on the production size batches."[3]

    "What we do want though, in addition to the life span estimated by small-scale, is that you confirm by heightened testing at manufacturing scale under a validation protocol."[104]

Most of the time, the vendor for the column resin typically has a CMC regulatory package available to address the following 3 concerns: (1) the quality control testing for resin release, (2) extractables from the resin and (3) effectiveness of cleaning and storage agents microbial control. If your Quality Unit doesn't have this package, request it.

Since most column resins, while product-dedicated, are reused, the validation must demonstrate the useful lifetime of the packed resin in the column. Small-scale validation runs are typically carried out and then compared to the results from the production-scale runs. An example of this approach is presented in Table 44.

As part of the validation lifecycle, a system needs to be in place to monitor column performance, so that if the column begins to degrade or perform outside the validated parameters, it can be immediately replaced or regenerated, as specified by appropriate SOPs and supported by process validation.

Table 44. Validation of column useful lifetime: Example of Rituxan® monoclonal antibody purification with Protein A affinity chromatographic process step[105]

Parameter Measured	Cycle < 20		Cycle > 200 (reused)	
	Small-Scale	Production-Scale	Small-Scale	Production-Scale
Product Yield (%)	98.8	97.1	100.5	97.4
Impurities (ppm) CHO Protein DNA	4956 4.7	3286 3.6	3200 1.7	949 < 0.2
Leaschables (ppm) Protein A	< 7.8	< 7.8	< 7.8	9.3
Chromatographic Profiles	Expected chromatographic shape	Comparable	Comparable	Comparable
Virus Removal ($\log_{10}$ clearance) X-MuLV SV40 MVM	2.5 1.4 1.5	Not Applicable	3.1 2.5 2.7	Not Applicable

4.3. Process Validation Concerns for a Filtration Step

There are several types of filtration methods used in the purification of biopharmaceuticals:

- Microfiltration (MF): uses small controlled pore sizes (typically 0.1 or 0.2 micron size) to physically separate the biopharmaceutical product from microbial and mycoplasmal contamination

- Nanofiltration (NF): uses small controlled pore sizes (typically 0.01 to 0.05 micron size) to physically separate the biopharmaceutical product from virus contamination.

- Ultrafiltration (UF): uses very small pore sizes (molecular cut off from 1 to 1,000 kilodaltons in size) to retain the biopharmaceutical product while allowing small molecules and buffer salts to pass through. UF is carried out in tangential flow mode, and is used either to concentrate the protein product or for buffer exchange (diafiltration)

The filtration parameters need to be set, and then validation is performed to determine the following:

- The intended purpose for the filtration

- Absence of impact on the product (not de-natured or aggregated)

- Leachables from the membrane filter into the product stream

The filter vendor typically has a CMC regulatory package available to address the quality control testing for filter release, the issue of extractables from filter membranes and biological safety testing (per USP Biological Toxicity of plastics) for the filter element components. If your Quality Unit doesn't have it, ask for it.

When MF and NF units, which are used to remove adventitious agents (microbes, mycoplasma, virus), are single-use and disposable. Therefore, cleaning validation is not required. However, the UF units contain single-use filter membranes but the filter holders (cassette) may be multiple-use. Cleaning a cross-flow filtration cassette can take up to a day, use lots of WFI and is required to be validated. Disposable cassettes for the UF unit are now available to reduce this cleaning validation cost.

5. IN-PROCESS CONTROLS

Throughout the purification process, in-process controls are expected at key process steps to provide an ongoing measure of the performance of the process and to ensure the safety of the purified product. At a minimum, the following quality control in-process tests should be performed:

- Step Yield (at selected process steps)

  During clinical development, expected yields can be more variable and less defined that the expected yields used in commercial processes[22]

  For commercial processes, actual yields should be compared with expected yields at designated steps in the production process; expected yields with appropriate ranges should be established based on previous laboratory, pilot scale, or manufacturing data[22]

- Bioburden (at selected process steps)

  Monitor the bioburden level throughout the manufacturing process to ensure the level is controlled[103]

- Endotoxin (at selected process steps)

  If there is a step in the manufacturing process to remove endotoxin, it is usually early in the process, so the remainder of the process must be carefully controlled against comtamination[103]

- Critical Process Impurities Related to Safety (at a selected process step)

  Antibotics used in cell culture (e.g., tetracycline, gentamicin)

  Serum residuals from cell culture media (e.g., BSA, bovine IgG); if serum is used at any stage, its calculated concentration in the drug product shall not exceed 1:1,000,000 (1 ppm)[29, 106]

  Immunogenic leachables from affinity column resins (e.g., Protein A- or monoclonal antibody-bound resins)[2, 107]

It is the expectation of the regulatory agencies that improvements in these process controls, as well as additional process controls, will be implemented by the manufacturer during the development of the process. For the BLA/NDA submission, not only is a complete description of all process controls expected, but also the justification of all in-process controls selected. Furthermore, the established in-process control test methods need to be validated for the BLA/NDA.

An example of the in-process controls for a FDA market approved purification process is presented in Table 45.

Table 45. In-process controls for the monoclonal antibody purification process, infliximab (Remicade®) [71]

Purification Process Step	In-Process Controls
Protein A Chromatography	pH Endotoxin by LAL Bioburden Protein concentration by OD_{280} Identity by IEF Purity by GF-HPLC
Solvent/Detergent Treatment	pH
Cation Exchange Chromatography	Endotoxin by LAL Purity by GF-HPLC

Nanofiltration	Protein concentration by OD$_{280}$
Anion Exchange Chromatography (Primary)	pH Endotoxin by LAL Purity by GF-HPLC
Anion Exchange Chromatography (Secondary)	Endotoxin by LAL Purity by GF-HPLC

6. PROCESS-RELATED IMPURITY PROFILE

Process-related impurities have three main origins in a biopharmaceutical manufacturing process:

- Cell Substrate

    Proteins derived from the recombinant host organism

    Nucleic acids (host cell genomic, vector or total DNA) derived from the recombinant host organism

- Cell Culture Media Components

    Inducers, antibiotics, serum

    Endotoxin

- Purification Reagents and Materials

    Enzymes, oxidizing/reducing agents, heavy metals, ligands (e.g., monoclonal antibodies), and other leachables

    Endotoxin

It is necessary to assure that these impurities are reduced to an acceptable level in the biopharmaceutical product. Three approaches are used to determine this:

- Retrospective analysis:

  Analysis of completed purification runs during clinical development (especially if the residual test was included in the in-process testing)

- Monitoring

  Measurement of residual levels at selected steps through the purification process (assuming that there are detectable levels after the first steps)

- Clearance studies

  Spiking experiments at small-scale to demonstrate the removal of a process-specific residual at a selected process step

At Phase 1. selected process-related residuals are typically tested at appropriate locations in the purification process or the final API. During later stage clinical development, the FDA expects clearance studies to be performed. There are four cases where clearance studies are expected to be performed:

- Antibiotics used in cell culture media[29]

- Host cell DNA[29, 108]

- Host cell protein[108]

- For any impurity claimed to be removed from a specific purification process step –

  "Manufacturers should have validation reports for the various key process steps. For example, if an ion-exchange column is used to remove endotoxins, there should be data documenting that this process is consistently effective. By determining endotoxin levels before and after processing, a manufacturer should be able to demonstrate the validity of this process. It is important to monitor the process before, during, and after to determine the efficiency of each key purification step. "Spiking" the preparation with a known amount of a contaminant to demonstrate its removal may be a useful method to validate the procedure."[3]

These clearance studies are performed in an attempt to identify the major purification steps capable of reducing the impurity burden and to document the capacity of those purification steps in reducing the impurity content in the product. Furthermore, the FDA indicates that the clearance studies may have to be repeated when manufacturing process changes are considered.

For the market authorization filings, both the FDA and the EMEA expect impurity testing on at least 3 consecutive production lots to confirm the removal of the process-related impurities for which clearance studies have been conducted.[29, 108]

The CMC review for FDA approval of the monoclonal antibody infliximab (Remicade®) illustrates one company's approach to process-related impurity profiling.[71] A three-fold approach was used: (1) direct measurement of residual in product, (2) monitoring removal of residual through the purification process, and (3) clearance studies. For each potential impurity, direct measurement of the API was performed (Table 46).

Table 46. Direct measurement of process-related residuals present in the purified monoclonal antibody, infliximab (Remicade®)[71]

Impurity	Measured Level in Product
Bovine Serum Albumin	< 0.4 ng/mg
Bovine IgG	0.8 - 14 ng/mg
Host Cell DNA	< 0.7 pg/mg
EDTA	< 0.15 mM
Pluronic F68	< 4.6 ug/mL
Excyte	-
Bovine Transferrin	0.2 - 1.2 ng/mg
Bovine Insulin	< 0.019 ng/mg
Mycophenolic Acid	< 15.5 ng/mL
Hypoxanthine	< 4.6 ng/mL
Xanthine	< 4.2 ng/mL
Hydrocortisone	< 0.3 ng/mL
Tween 80	< 20 ug/mL
Tri-n-Butyl Phosphate	< 0.5 ug/mL
Protein A	< 125 ng/mg
Guanidine	< 5.3 ug/mL

The removal of one impurity (bovine IgG) was monitored through the purification process (Table 47) and it showed that this impurity was primarily removed from the product at the first chromatographic step. Clearance studies were performed on 5 impurities (BSA, host cell DNA, Excyte, Protein A and guanidine) to demonstrate several logs of removal by subsequent chromatography steps.

Table 47. The removal bovine IgG impurity through the purification process of the monoclonal antibody, infliximab (Remicade®) (results in ng per mg of MAb)[71]

Purification Process Step	Lot 1	Lot 2	Lot 3	Lot 4
Protein A	15	18	7	102
Cation Exchange	9	10	3	87
Primary Anion Exchange	7	10	3	87
Secondary Anion Exchange	2	2	1	14

Based on this three-fold approach to study process-related impurity removal, only two impurities were of potential concerns: mycophenolic acid and bovine IgG. Mycophenolic acid is used for selection of gpt expressing C168J transfectants during the cell culturing process. However, mycophenolic acid is also a poison. Centocor was able to effectively argue that the highest theoretical residual level of mycophenolic acid in infliximab was lower that the licensed dose for mycophenolate mofetil, which is used for the prevention of transplanted organ rejection. Since this was acceptable to the FDA, they did not need to perform any additional studies. Bovine IgG (from the BSA used in the cell culture process) is immunogenic. Centocor was required to set a specification for bovine IgG, but not until 10 lots had been manufactured at the commercial scale.[71]

7. VIRAL SAFETY EVALUATION

Prevention of virus contamination is achieved through incorporation of various barriers placed in the manufacturing process. Choices made by the manufacturer – concerning the selection of cell lines not containing endogenous viruses, the acceptance criteria of any animal-derived materials used in production (sourcing policy, vendor qualification, virus testing), the rigorous enforcement of cGMPs during production, and the testing of each production harvest lot for the absence of virus – all contribute to ensuring that viruses do not contaminate the biopharmaceutical product. But there is one additional regulatory control: the need to assess the capacity of the purification process to clear infectious viruses. This assessment is done by performing viral clearance studies.

It should be noted, that while animal- and human-derived raw materials are frequently used in the production process, they also are occasionally found in the purification process. For example, the Protein A used in affinity purification columns is purified over an immobilized human immunoglobulin resin, thus exposing the Protein A to potential human viruses. For this reason, many manufacturers are now switching to recombinant Protein A, which avoids this problem. Another example is the use of Tween 80 (polysorbate 80) as a detergent in the purification process. Tween 80 is produced from fatty acids which come from bovine tallow. Manufacturers now have the choice of using Tween 80 manufactured from plant-derived fatty acids. Animal- and human-derived materials should be avoided if at all possible because of the risk of introducing viruses and TSEs.

Viral clearance studies serve two purposes:

1. Evaluation: to demonstrate the ability of the purification process to clear any virus known to be present in the production harvest material

2. Characterization: to assess the robustness of the purification process to clear adventitious viruses that could possibly contaminate the production harvest material

These studies are the 'final safety net' for protecting the product. By demonstrating excess capacity for viral clearance built into the purification process, a manufacturer can assure an appropriate level of safety

7.1. General Study Design

Viral clearance studies need to be designed, conducted and analyzed in a manner that provides accurate information to reliably assess the ability of the purification process to clear viruses. Table 48 presents some recommendations from regulatory authorities for carrying out these studies.

Table 48. Some points to consider when carrying out viral clearance studies[29, 69]

Facilities and Staff	Viruses are not allowed in a manufacturing facility. Therefore, virus clearance studies should be conducted in a separate laboratory equipped for virological work and performed by staff with virological expertise in conjunction with manufacturing personnel involved in designing and preparing the scaled-down version of the purification process.
Scaled-Down Purification Process	The validity of the scaling down must be demonstrated. The level of purification of the scaled-down version should be representative of the purification process performed in manufacturing. Deviations which cannot be avoided due to scaling down must be discussed and their influence on the results assessed.

Step-Wise Elimination of Virus	The contribution of more than one purification step must be assessed. Sufficient virus should be present in the material of each step to be tested so that an adequate assessment of the effectiveness of each step is obtained. Generally, virus should be added to in-process material of each step to be tested. The virus titer before and after each step needs to be tested.
	Both virus removal and virus inactivation must be studied.
Function and Regeneration of Columns	Over time and after repeated use, the ability of chromatography columns and other devices used in the purification scheme to clear virus may vary. Some estimate of the stability of the viral clearance after several uses provides support for repeated use of such columns. Assurance should be provided that any virus potentially retained by the purification system would be adequately destroyed or removed prior to reuse of the system.
Specific Precautions	Care should be taken in preparing the high-titer virus to avoid aggregation which may enhance physical removal and decrease inactivation thus distorting the correlation with actual manufacturing.
	Consideration should be given to the minimum quantity of virus which can be reliably assayed.
	The study should include parallel control assays to assess the loss of infectivity of the virus due to such reasons as the dilution, concentration, filtration or storage of samples before titration.
	The virus 'spike' should be added to the product in a small volume so as not to dilute or change the characteristics of the product. Diluted, test-protein sample is no longer identical to the product obtained in manufacturing.
	Virus inactivation is time-dependent, therefore, the amount of time a spiked product remains in a particular buffer solution or on a particular chromatography column should reflect the conditions of the purification process.
	Buffers and product should be evaluated independently for toxicity or interference in assays used to determine the virus titer, as these components may adversely affect the indicator cells. If the solutions are toxic to the indicator cells, dilution, adjustment of the pH, or dialysis of the buffer containing spiked virus might be necessary. Sufficient controls to demonstrate the effect of procedures used solely to prepare the sample for assay (e.g., dialysis, storage) on the removal/inactivation of the spiked virus should be included.
	An effective virus removal step should give reproducible reduction of virus load shown by at least two independent studies.
Interpretation of Results	Virus infectivity assays used to quantitate the virus titer should be sensitive, reproducible and conducted with sufficient replicates to demonstrate statistical accuracy.
	Assays used must have known between-assay precision. Results of the reference preparation must be within approximately $0.5 \log_{10}$ of the mean estimate established in the laboratory for the assay to be acceptable.
	The 95% confidence limits for test results of within-assay variation normally should be on the order of $\pm \log_{10}$ of the mean.

Reduction in virus tier of the order of 1 $\log_{10}$ or less is negligible and should not be included in the overall reduction factor calculation.

Addition of individual virus reduction factors resulting from similar inactivation mechanisms along the purification process may overestimate viral clearance.

Viruses are cleared by two different mechanisms:

1. Removal of the virus (e.g., chromatography, nanofiltration, etc.)

2. Inactivation of the virus (e.g., pH treatment, heat, detergent treatment, etc.)

The FDA recommends that the purification process be designed to include at least two orthogonal robust viral clearance steps.[9] Viral clearance is the sum of virus removal and virus inactivation.

7.2. Justification of the Choice of Viruses

A critical component of the design of a viral clearance study is the justification of the choice of viruses to use. Viruses should be chosen to resemble viruses which may contaminate the product and to represent a wide range of physicochemical properties in order to test the ability of the purification process to eliminate viruses in general:

- 'Relevant' viruses – those which are either the identified viruses, or of the same species as the viruses, that are known, or likely to contaminate the production process

  Examples:. the use of a murine retrovirus for murine cell lines (e.g., NS0) known to contain infectious endogenous retroviruses; the use of bovine viral diarrhea virus (BVDV) for cell lines contaminated with hepatitis C virus (HCV)

- 'Specific model' viruses – those which are closely related to the known or suspected virus (same genus or family), having similar physical and chemical properties to the observed or suspected viruses

  Examples: the use of xenotrophic murine leukemia virus (XMuLV) for cell lines derived from rodents that contain endogenous non-infectious retrovirus-like particles, like Chinese hamster ovary (CHO); the use of herpes simplex virus (HSV) or pseudorabies virus (PRV) for human cell lines immortalized by Epstein-Barr virus (EBV)

- 'Non-specific model' viruses – those which encompass a wide range of physico-chemical properties (size, enveloped capsid, and resistance to heat and chemical treatment) to demonstrate the robustness of the purification process for clearing viruses

   Characterization of the viral clearance capability of the purification process typically requires testing the removal/inactivation of at least 4 non-specific model virus types (Table 49).

Table 49. Some non-specific model viruses used in virus clearance studies

Genome	Enveloped*	Diameter	Resistance to Physicochemical Treatments**	Examples
DNA	No	≤ 50 nm	Very High	Murine Minute Virus (MMV) Porcine parvovirus (PPV)
RNA	No	20-80 nm	Medium to High	Reovirus (Reo) Hepatitis A (HAV)
DNA	Yes	≥ 120 nm	Medium	Herpes simplex virus (HSV) Pseudorabies virus (PRV)
RNA	Yes	50-110 nm	Low	Murine leukemia virus (MuLV) Xenotropic murine leukemia virus (XMuLV)

* Enveloped refers to a lipid bilayer surrounding the virus capsid
** Heat, high/low pH, solvents, detergents (e.g., Triton X-100)

7.3. Calculation of Virus Reduction Factors

For each purification process step, a virus reduction factor can be calculated, which is defined as the $\log_{10}$ of the ratio of the virus load (i.e., titer x volume) in the pre-purification material and that in the post-purification material. Some typical virus reduction factors expected by the FDA include[29]:

Low pH treatment (pH 3.9)	3 - 4 $\log_{10}$ inactivation
Heat	$\sim$ 4 $\log_{10}$ inactivation
Nanofiltration (15-40 nm pore size)	4 - 8 $\log_{10}$ removal

All purification process steps do not need to be tested, but it is required that at least one virus inactivation step and one virus removal step be tested.

An overall virus reduction factor is calculated from the sum of the individual $\log_{10}$ reduction factors. This value represents the $\log_{10}$ ratio of the virus load at the beginning of the purification process and at the end of the purification process. To prevent an overestimation of the overall virus clearance by the purification process, repetition of identical or near identical virus clearance mechanisms can not be included in the overall virus reduction factor. For example, the total viral clearance from the use of a Protein A affinity chromatography step is the sum of virus removal due to lack of binding by the virus to the resin plus the virus inactivation due to contact of the virus with the low pH buffers used on the column. If it was desired to add the viral clearance results from a subsequent low pH treatment step further down the purification process to the overall virus reduction factor, the contribution of the low pH buffers to the virus inactivation on the Protein A column would need to be factored out. Also, only individual virus reduction factors greater than 1 can be counted in the overall virus reduction calculation.[69]

An example of a viral clearance study for a commercial purification process is presented in Table 50.

Table 50. Viral clearance results from the purification process for recombinant human coagulation Factor VIIa (NovoSeven®)[109]

Virus Tested	Murine leukemia virus (MuLV)	Reovirus type 3 (Reo-3)	Infectious bovine rhinotracheitis virus (IBR)	Bovine enterovirus (BEV)	Simian virus 40 (SV40)
Virus Type	RNA env	RNA non-env	DNA env	RNA non-env	DNA non-env
Purification Step					
Chromatography Step 1	4.7	7.6	4.5	4.8	3.6
Chromatography Step 2	> 6.1	Not tested	> 4.7	0.4	Not tested
Chromatography Step 3	2.4	5.9	5.0	3.4	2.8
Detergent Treatment	> 6.3	> 9.3	2.9	> 8.0	> 3.0
Overall Virus Reduction Value	> 19.5	> 22.8	> 17.1	> 16.2	> 9.4

Many variables affecting viral clearance are well understood, but the actual virus reduction factors obtained by a manufacturer are influenced by the virus strain selected, the variability of the assay procedures performed and the nature of the product itself being purified. For example, the pH inactivation of murine leukemia virus (MuLV) ranged from 1.7 to 7.0 $\log_{10}$ reduction values at pH 3.7, and ranged from 4.1 to 7.1 at pH <3.0, for different products tested by different manufacturers. MuLV in protein solution at the same pH was either completely inactivated by 60 minutes, or by 120 minutes, or even greater than 120 minutes, depending upon the manufacturer who performed the test.[110] It is for this reason that caution needs to be exercised in trying to predict virus clearance factors from published reports. However, generic and modular virus clearance studies that could be used to predict virus clearance within a manufacturer's range of their own products are being pursued as a means of ultimately saving considerable time and cost. The FDA has indicated a willingness to evaluate these approaches.[29]

7.4. Virus Safety Calculation

The end result of the viral clearance studies is for the manufacturer to be able to estimate the risk of virus particles in doses of product administered. From the calculated overall virus reduction factor of the relevant and/or specific model virus(es), the viral safety calculation is calculated. Table 51 presents the calculation. In the example provided, less than one virus particle per million doses of product would be expected. This viral safety calculation needs to be performed only on the relevant and/or specific model virus(es).

Table 51. Example of the calculation of estimated virus particles per dose

Three values are needed for this calculation:

1. Measured or estimated concentration of virus in production harvest

   Typically, retrovirus particles are quantitated using transmission electron microscopy (for this calculation we will use 10^8 particles/mL)

2. Calculated overall viral clearance factor

   Typically, the specific model virus is chosen for the calculation (for this calculation we will use $>10^{17}$ $\log_{10}$)

3. Volume of production harvest needed to purify and make a dose of product (for this calculation we will use 1000 mL/dose)

The calculation of the estimated virus particles per dose is as follows:

$$(1) \ X \ (3) \ / \ (2) = \ < 10^6 \text{ particles per dose}$$

What viral clearance safety factor are the regulatory agencies expecting to see? According to the regulatory agencies, they have not officially set a safety factor number:

"These studies are not performed to evaluate a specific risk factor. Therefore, a specific clearance value needs not to be achieved."[69]

However, the FDA expects that the purification process will be able to remove or inactivate retroviruses in excess of the endogenous particle load.[29]

7.5. Worth All the Trouble and Cost?

If there was one area where senior management questions the expenses, it is in the cost of completing these viral clearance studies. Expenses readily exceeding hundreds of thousands of dollars are common to complete the full panel of non-specific virus types. However, these studies are not to assure that known viruses are cleared by the purification process, but instead are performed to protect against putative viruses should they ever be present. So it is reasonable to question if it is worth all of the expense and resource commitments to complete these studies.

The recent virus scares impacting our blood supply answers the question. The West Nile Virus (WNV) is a small (50 nm) lipid-enveloped RNA virus[24]; the causative agents for the Severe Acute Respiratory Syndrome are also lipid-enveloped RNA viruses[25]; the live vaccinia virus in the smallpox vaccine is a lipid-enveloped DNA virus[111]; while the monkeypox virus is a virus related to vaccinia.[112] Because of the thorough viral clearance studies carried out by the human plasma product manufacturers, the FDA was able to assure the public that the plasma-derived components, some of which are used in biopharmaceutical manufacturing, are not posing a risk to patients. The following is the statement made concerning SARS:

> "At this time, we believe SARS is unlikely to be transmitted through products manufactured from plasma. Lipid-enveloped RNA virus(es), the putative agent(s), should be readily removed and/or inactivated during manufacturing of plasma derivates. Licensed plasma derivatives undergo intentional viral clearance procedures that are validated to be effective against lipid-enveloped RNA viruses. These procedures include: filtration, heating, acidification, and detergent treatment."[25]

The biopharmaceutical manufacturer can take some comfort in knowing that their viral clearance studies do provide an extract layer of safety over their purification process.

7.6. When Should the Viral Clearance Studies Be Performed?

It is clear that the full viral clearance studies must be completed for the BLA/NDA filing. The omission of data demonstrating the absence or removal of adventitious agents is consider a CMC reason for refusal to file a BLA/NDA.[31]

But how much of this needs to be done at Phase 1 and the early stages of clinical development? According to the FDA:

> "Retrovirus clearance studies should be performed prior to Phase 1 trials .... Clearance studies for other viruses should be carried out prior to production for Phase 2/3 trials and may need to be repeated if the final manufacturing process has changed."[29]

For many biopharmaceutical processes, a single virus clearance study would be sufficient. However, if the production system uses a novel host or a host that contains an endogenous virus, this would not be sufficient. The FDA will place a biopharmaceutical product on Phase 1 IND clinical hold if they feel that there is an inadequate demonstration of viral clearance.[39]

8. PURIFICATION CONTROLS FOR GENE THERAPY PROCESSES

The gene therapy biopharmaceuticals are DNA-based and not protein-based. Therefore, rather than trying to purify DNA impurities away from the expressed protein, the challenge of the purification process is to purify protein impurities away from the expressed DNA. It is therefore no surprise that chromatographic purification of the DNA plasmids and viral vectors is primarily with anion exchange chromatography (including the membrane-based systems).

While the purification processes for these products will be different, the same CMC regulatory issues will apply (e.g., application of cGMPs, detection of adventitious agents, validation of the purification process, etc.). Process-related impurity removal needs to be addressed (e.g., RNA, endotoxin, proteins, DNA isoforms, antibiotics, etc.). Validation of the purification process will need to be completed.

As these biopharmaceuticals move toward the later clinical development stages, expect to see more CMC regulatory guidance being provided for these purification processes.

9. WHAT CAN GO WRONG

With all of the things that need to be properly controlled during purification of a biopharmaceutical, companies in their haste can fail to meet the requirements expected by the regulatory agencies. This usually results in inspectional issues when the FDA arrives at the manufacturing facility. Table 52 presents some purification process control deficiencies recorded by the FDA.

Table 52. Purification process deficiencies recorded by FDA inspectors[113, 114, 115]

FDA Establishment Inspection Report (EIR) for MedImmune, Inc., March 1998

The Manufacturing and the Quality Unit approved commercial scale purifications based on scaled down viral clearance studies for the columns. Their review did not identify differences in the operating parameters and maximum protein load per unit resin between the scaled down column and the commercial manufacturing columns. I expressed my concern that the Quality Unit reviewed and approved the batch records when, in fact, some operating parameters and process buffers were not the same as what was used in the viral clearance studies.

I reviewed the cleaning validation for the equipment used to manufacture palivizumab. I noted that the cleaning validation study included a hold time for used tanks prior to cleaning, however, in manufacturing, used tanks may be held up for a longer period of time prior to cleaning. I asked Dr. Tsao why the maximum hold time used in manufacturing was not included in the cleaning validation study. Dr. Tsao replied that over the time the need to hold vessels longer increased, but they did not revalidate the hold time.

We recommend replacement of difficult to clean surfaces, such as hanging ceiling tiles and floor ledges, in the cell culture and purification suites with hard smooth cleanable surfaces.

Filter extractable studies have not been performed for the filters used for the in-process palivizumab bulks and buffers. These filters are not flushed to remove potential filter extractables prior to use.

FDA Warning Letter to IDEC Pharmaceuticals, November 1999

Please provide additional information that demonstrates that the implicated lots of product do not contain elevated levels of endotoxin. Please note that we are aware that endotoxin removal has not been evaluated in your purification process

We note that IDEC Pharmaceutical Corporation performs an LAL at process step – Cell Free Receiving Tank, and reports the result as a 'Report Value.' Please explain the purpose of a 'Report Value' including how Quality Assurance and Production managers use this information to assess product quality and purity. In addition, please explain the benefit of collecting this information if atypical results (when compared to historical data) are not investigated, i.e., an assessment of impact to product and other associated batches, identification of the cause of the problem, and implementation of corrective actions.

FDA Establishment Inspection Report (EIR) for Genentech, August 2000

For Pulmozyme, removal of retrovirus was submitted in the PLA. MVM removal and inactivation had been submitted as an amendment to the PLA. For Rituxan, Herceptin and TNKase, I verified that similar studies had been performed demonstrating clearance of at least one other virus besides retrovirus for each process/product, and these studies were also performed as part of the column reuse validation. I discussed with Dr. Xu the current ICH and PTC guidelines regarding viral clearance studies, and emphasized the fact that current recommendations suggest the use of 3 model viruses for each clearance study, and viral clearance studies are an important component of column reuse validation. Dr. Xu voiced her opinion at having received different pieces of advice when contacting the Agency regarding updating these types of studies. I made it clear to her and to the attendees during the end-of-day discussion that regardless of what anyone might say or conflicting advice one might receive, the PTC guidelines and ICH recommendations describe the minimum requirements.

There are no validation studies performed to demonstrate that in-process buffers and resin used in the rhuMAb HER2 purification process is non reactive to 316L stainless steel columns (i.e., Column C1130).

During the inspection, I stated I was concerned that manufacturing could repeat processing steps, change agitation rates and times, and also use expired bulks and buffers that did not meet specifications, all of which

are failure to follow the batch record. In addition, I stated I was concerned at the lack of Quality oversight into these events and that any review by Quality happens well after the event.

Equipment used to manufacture product is not constructed so that the product contact surfaces are non-reactive. For example, during the Herceptin chromatography operations, discoloration of the resin and corrosions was observed on the top head plate, column body and the bottom flow adapter.

10. STRATEGIC CMC TIPS FOR PURIFICATION

The purification process for a biopharmaceutical at an early clinical stage will be basic and somewhat unpolished. The focus needs to be on ensuring safety for the product to be purified, as indicated by Figure 12.

Figure 12. Minimum CMC continuum for a biopharmaceutical purification process

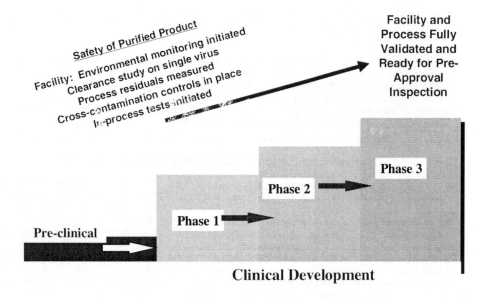

While the list in the Figure will vary with the type of biopharmaceutical being purified, the recombinant host used in the production system, the complexity of the facility operations, and even the person drawing up the list, each of the activities selected should be related to ensuring that the expressed biopharmaceutical is safe for the patient.

The list of CMC 'must haves' increases as the clinical development moves into Phase 2 and Phase 3. The focus now needs to include purification process control and robustness. Critical purification parameters need to be identified, operating ranges set, viral clearance studies completed, and the process determined to be robust enough to tolerate the manufacturing environment and produce marketable product.

By time of the BLA/NDA filing, all CMC regulatory compliance issues related to the purification process are to be completed.

I reviewed the cleaning validation for the equipment used to manufacture palivizumab. I noted that the cleaning validation study included a hold time for used tanks prior to cleaning, however, in manufacturing, used tanks may be held up for a longer period of time prior to cleaning. I asked Dr. Tsao why the maximum hold time used in manufacturing was not included in the cleaning validation study. Dr. Tsao replied that over the time the need to hold vessels longer increased, but they did not revalidate the hold time.

We recommend replacement of difficult to clean surfaces, such as hanging ceiling tiles and floor ledges, in the cell culture and purification suites with hard smooth cleanable surfaces.

Filter extractable studies have not been performed for the filters used for the in-process palivizumab bulks and buffers. These filters are not flushed to remove potential filter extractables prior to use.

FDA Warning Letter to IDEC Pharmaceuticals, November 1999

Please provide additional information that demonstrates that the implicated lots of product do not contain elevated levels of endotoxin. Please note that we are aware that endotoxin removal has not been evaluated in your purification process

We note that IDEC Pharmaceutical Corporation performs an LAL at process step – Cell Free Receiving Tank, and reports the result as a 'Report Value.' Please explain the purpose of a 'Report Value' including how Quality Assurance and Production managers use this information to assess product quality and purity. In addition, please explain the benefit of collecting this information if atypical results (when compared to historical data) are not investigated, i.e., an assessment of impact to product and other associated batches, identification of the cause of the problem, and implementation of corrective actions.

FDA Establishment Inspection Report (EIR) for Genentech, August 2000

For Pulmozyme, removal of retrovirus was submitted in the PLA. MVM removal and inactivation had been submitted as an amendment to the PLA. For Rituxan, Herceptin and TNKase, I verified that similar studies had been performed demonstrating clearance of at least one other virus besides retrovirus for each process/product, and these studies were also performed as part of the column reuse validation. I discussed with Dr. Xu the current ICH and PTC guidelines regarding viral clearance studies, and emphasized the fact that current recommendations suggest the use of 3 model viruses for each clearance study, and viral clearance studies are an important component of column reuse validation. Dr. Xu voiced her opinion at having received different pieces of advice when contacting the Agency regarding updating these types of studies. I made it clear to her and to the attendees during the end-of-day discussion that regardless of what anyone might say or conflicting advice one might receive, the PTC guidelines and ICH recommendations describe the minimum requirements.

There are no validation studies performed to demonstrate that in-process buffers and resin used in the rhuMAb HER2 purification process is non reactive to 316L stainless steel columns (i.e., Column C1130).

During the inspection, I stated I was concerned that manufacturing could repeat processing steps, change agitation rates and times, and also use expired bulks and buffers that did not meet specifications, all of which

are failure to follow the batch record. In addition, I stated I was concerned at the lack of Quality oversight into these events and that any review by Quality happens well after the event.

Equipment used to manufacture product is not constructed so that the product contact surfaces are non-reactive. For example, during the Herceptin chromatography operations, discoloration of the resin and corrosions was observed on the top head plate, column body and the bottom flow adapter.

Biopharmaceutical Drug Product Manufacturing

"The filling of biotechnology-derived products (BDP) into ampoules or vials presents many of the same problems as with the processing of conventional products. In established companies these issues are relatively routine. However, for the new BDP facility, attempting to develop and prove clinical effectiveness and safety along with validation of sterile operations, equipment and systems, can be a lengthy process, particularly if requirements are not clearly understood."

1991 FDA biotechnology inspection guide[3]

1. THREE BASIC CMC REGULATORY CONCERNS

At this point in the overall manufacturing process, the production system used to express the biopharmaceutical or the purification process used to yield the purified active pharmaceutical ingredient (API), matters little. The API, regardless of its source and handling, at this point is either a pure recombinant protein or a pure DNA material.

The purified API typically needs to be formulated with inactive ingredients, and then filled into some type of suitable container and sealed. This is the drug product. Following this, labels are placed on the drug product containers and then the labelled drug product is packaged for patient administration.

Regulatory agencies have expressed three major concerns about biopharmaceutical drug product manufacturing:

1. Control of the Aseptic Process

    The greatest risk for the biopharmaceutical comes from the type of manufacturing process that has to be employed: aseptic processing.

    Terminal sterilization, which is used for many chemically-synthesized drug products, usually involves filling and sealing product containers under conditions of a high quality environment; the product, container, and closure in most cases have low bioburden but are not sterile. The environment in which filling and sealing is performed is of high quality in order to minimize the microbial content of the in-process product, and to help ensure that the subsequent sterilization process is successful. The product in its final container is then subjected to a sterilization process such as heat or radiation.

    In aseptic processing, which has to be used for biopharmaceuticals because they are thermally labile, the drug product, container, and closure are subjected to sterilization processes separately and then brought together. Because there is no further processing to sterilize the product after it is in its final container, it is critical that containers be filled and sealed in an environment of extremely high quality.

    Failure to stringently adhere to cGMPs during the drug product manufacturing of a biopharmaceutical can impact product safety and efficacy. A contaminated drug product poses an unacceptable risk to a patient.

2. Selection of Excipients for the Formulaiton

    According to 21 CFR 210.3(b)(8), an inactive ingredient is any component of a drug product other than the active ingredient.[116] These inactive ingredients are called excipients. For most biopharmaceuticals, there is more excipient mass than active drug mass in the drug product; thus, a manufacturer has to consider the safety risks of these inactive ingredients.

3. Consistent Drug Product Manufacture

    The manufacturing process must have adequate procedures and controls in place to ensure that the process is consistently performed and that it yields a consistent product of required quality and safety.

    The formulated API is filled into containers under aseptic process conditions. If the drug product is to be a liquid then either suitable closures are placed onto the containers to seal the product (e.g., rubber stoppers on glass vials or syringes) or the containers are sealed directly (e.g., ampoules, plastic bags, tubes).   If the drug product is to be a solid, the filled containers are placed in a freeze-dryer first, lyophilized, and then suitable closures are placed onto the containers to seal the product.

It is the responsibility of each manufacturer to demonstrate to the regulatory agencies that their drug product manufacturing process has been designed for product safety and that it is adequately controlled and monitored.

A manufacturer does not really have a useable drug product manufacturing process until it can be shown that the process can be consecutively performed, and that it can yield a consistent product of required quality.   Unlike the production and purification stages of biopharmaceutical manufacturing where continual development and improvement of the process is expected during clinical development, changes in drug product manufacturing will be more limited. Because this process prepares the product that will be administered to patients, adherence to cGMP and achieving the necessary quality controls and product safety requirements at the outset is required.   The main development and improvement should be in the formulation or drug delivery designs.

2. FORMULATION OF A BIOPHARMACEUTICAL

Pharmaceutical excipients are any substance, other than the active drug, which has been appropriately evaluated for safety and is included in a drug delivery system to either:

- Aid processing of the system during manufacture

- Protect, support or enhance stability, bioavailability, or patient acceptability

- Enhance any other attribute of the overall safety and effectiveness of the drug product during storage and use

2.2. Formulations Familiar to FDA

These excipients, while inactive ingredients, may not be harmless. Adulteration of excipients has caused significant safety issues with some chemically-synthesized drug products. For example, in 1996, an epidemic of acute renal failure in Haiti was caused by acetaminophen manufactured in that country with the excipient glycerine. The glycerine which may have originated in China was contaminated with toxic diethylene glycol.[117]

The FDA expects the manufacturer to be able to justify the following four requirements for any excipients that are used:

- Safe for its intended use

- Pure

- Well-defined identity and source

- High quality

One key safety assessment for an excipient under consideration is its prior use in the formulation of a marketed biopharmaceutical drug product. Table 53 presents a list of some excipients used in market approved biopharmaceuticals.

Table 53. Some excipients used in the formulation of market approved biopharmaceuticals; listed on the FDA website: www.fda.gov/cber/prodcuts.htm

Biopharmaceutical	Formulation Components
Palivizumab, Synagis® Monoclonal Antibody	Histidine, glycine, mannitol
Infliximab, Remicade® Monoclonal Antibody	Sucrose, polysorbate 80, sodium phosphate monobasic, sodium phosphate dibasic
Basiliximab, Simulect® Monoclonal Antibody	Mannitol, glycine, sucrose, sodium chloride, potassium phosphate monobasic, disodium hydrogen phosphate
Interferon Beta-1a Rebif®	Albumin (human), mannitol, sodium acetate
Coagulation Factor VIIa NovoSeven®	Mannitol, sodium chloride, calcium chloride, glycylglycine, polysorbate 80
Interleukin-2 fused to diphtheria toxin ONTAK®	Citric acid, EDTA, polysorbate 20
TNF Receptor Fc Embrel®	Mannitol, sucrose, tromethamine
Trastuzumab, Herceptin® Monoclonal Antibody	Histidine, trehalose, polysorbate 20
Platelet-Derived Growth Factor-BB Regranex® Gel	Sodium carboxymethylcellulose, sodium chloride, sodium acetate, acetic acid, methylparaben, propylparaben, m-cresol

If a manufacturer wants to introduce a novel excipient into their formulation, additional resources must be provided to address FDA's concerns. A manufacturer not only has to convince the FDA that their biopharmaceutical product is safe but also that the novel excipient is safe. Toxicology data, either existing in the literature or needing to be generated by the manufacturer, will be needed. Expect the FDA to be conservative in their approach to something new:

> "Novel excipients are those that are used in the United States for the first time in a human drug product or by a new route of administration. The manufacturing, chemistry, and controls (CMC) information for a novel excipient should be provided in the same level of detail as that provided for a drug substance."[42]

> "It is important to perform risk-benefit assessments on proposed new excipients in drug products and to establish permissible and safe limits for these compounds. Safety data should be submitted to support use of new excipients."[118]

The FDA has prepared a searchable database to obtain information on the use of inactive ingredients in approved drug products.[119] The database provides information on the route of administration, dosage form and amount of excipient present in approved drug products. The database can aid in the selection of excipients to be considered for a biopharmaceutical formulation. However, it should be cautioned that not all excipients found in approved biopharmaceuticals appear to be listed in the database (e.g., trehalose is not present; to locate TRIS one must use tromethamine).

2.3. Chemcial Modification of API Prior to Formulation

To improve the properties of the biopharmaceutical product, chemical modification of the bulk API prior to formulation is also possible. Pegylation of proteins (i.e., the chemical attachment of polyethylene glycol polymers (PEG) through either aldehyde or succinimide ester or maleimide linkers to proteins) has been successfully applied. Some of the possible advantages of pegylation are:

- Increased half-life in vivo

- Decreased immunogenicity

- Increased physical and thermal stability

- Increased solubility

Pegylation is a chemical reaction, so a manufacturer must exercise careful control over the quality of the PEG raw material (range of molecular weight species), the quality of the PEG-linker (avoidance of bi-functional PEGs) and the reaction conditions (to control the site and extent of pegylation).

Several pegylated biopharmaceuticals have been approved by the FDA and appear on their website (www.fda.gov/cber/products.htm):

- Neulasta®, pegfilgrastim (granulocyte colony stimulating factor)

- Pegasys®, peginterferon alfa-2a

- PEG-Intron®, peginterferon alfa-2b

3. BIOPHARMACEUTIAL MANUFACTURING PROCESSES

Regulatory agencies, since they review and approve the drug product manufacturing process to grant market approval, are very familiar with numerous processes. Table 54 presents some examples of commercial drug manufacturing processes for biopharmaceuticals.

Table 54. Drug product manufacturing information from FDA approved biopharmaceuticals[63, 120]

Rituximab (Rituxan®) Monoclonal Antibody

Rituximab is received from IDEC as a formulated bulk that has been filtered into pre-sterilized bags and shipped from IDEC to Genentech. The product is filtered into a steam-sterilized stainless steel freeze/thaw tank. The product is sampled for endotoxin and bioburden prior to filtration and protein concentration after filtration. The filtered bulk may be frozen if required, hold times and number of freeze/thaw cycles has been validated. The hold tanks are pooled to give the desired batch size, the pooled bulk may be held for 48 hours at 2-8°C. The product is filled as either 10cc or 50cc sterile, depyrogenated USP Type I borosilicate glass vials with 20mm gray butyl rubber liquid stoppers and sealed with a 20mm aluminum/plastic flip-off type cap. The process gas is Nitrogen. Vials are then stored for inspection and release by quality assurance and quality control. Filled vials are 100% manually inspected prior to labelling. Vials are then labelled and released for shipping after release by quality assurance.

Palivizumab (Synagis®) Monoclonal Antibody

The purified product is formulated in 25 mM histidine, 1.2 mM glycine, 3% mannitol, pH 6.0 buffer. The resulting formulated bulk is 0.2 micron filtered into 5-L or 7-L Stedim 38 film, BioPharm XL bags and stored at 2-8°C until final filtration, fill, and lyophilization at an offsite contract manufacturer.

The container/closure system for lyophilized vialed product consists of 5 cc Schott, USP/EP type I borosilicate (Purform) glass vials with a 20 mm neck opening, 20 mm 4405/50 gray butyl rubber stoppers, and 20 mm flip-off aluminum overseals with a green button.

At the contract manufacturing site, sterile filtration occurs in a HEPA filtered Class 100 hood within the filling core. The bulk is passed through a 0.2 micron filter using nitrogen overpressure into a sterilized bulk tank located in the Aseptic Filling Room. Pre and post use integrity testing of the filter is performed.

For filling operations, the components are prepared and sterilized according to validated procedures for the vial washer, autoclave and depyrogenation oven. Product contact items and filling machine parts are dedicated to the product and sterilized according to validated procedures. Sterilized stoppers are loaded into the stopper hopper. The sterilized filling syringe and valve assembly are installed in the filling machine using sterilized tubing. Fill volume checks are performed and any necessary adjustment made. The vials are then filled with sterilized bulk, a stopper is partially inserted into each vial, and the filled vials are discharged into lyophilizer trays.

As each lyophilizer tray is filled with partially stoppered product vials, it is loaded into the prechilled freeze dryer. After all the filled vials have been loaded into the lyophilizer, the door is secured and the freeze drying cycle is begun. At the end of the lyophilization cycle, 0.2 micron filtered medical grade nitrogen is bled into the chamber. The stoppers are fully inserted into the vials by hydraulically raising the lyophilizer shelves.

After the stoppers are fully inserted into the vials, the vials are removed from the lyophilizer. Vials from each lyophilizer tray are loaded onto the accumulator in the Aseptic Filling Room and fed out to the oversealer located in the Finishing Room.

A variety of container/closures systems and drug delivery approaches have also been successfully applied to biopharmaceuticals (Table 55).

Table 55. Drug delivery formats from FDA approved biopharmaceuticals; listed on the FDA website: www.fda.gov/cber/prodcuts.htm

Physical Form	Container/Closure System	Application Route	Manufacturing Information
Liquid Product	Vial	Injection	Rituxan®, Chimeric Monoclonal Antibody
			Rituxan®, a sterile, clear, colorless, preservative-free liquid in single-use vials for intravenous (IV) administration

Liquid Product	Ampoule	Inhalation	Pulmozyme®, Dornase Alfa
			Pulmozyme, a sterile, clear, colorless solution of dornase alfa. It is supplied in single-use ampoules. It is administered by inhalation of an aerosol mist produced by a compressed air driven nebulizer system.
Liquid Product	Syringe	Injection	Rebif®, Interferon Beta-1a
			Rebif® is formulated as a sterile solution in a prefilled syringe intended for subcutaneous (SC) injection.
Lyophilized Product	Vial	Injection	Proleukin®, Interleukin 2
			Proleukin® is supplied as a lyophilized solid in single-use vials for IV infusion
Solid Product Embedded in Biodegradable Microspheres	Vial	Injection	Nutropin Depot®, Somatropin, Human Growth Hormone
			Nutropin Depot formulation consists of micronized particles of rhGH embedded in biocompatible, biodegradable polyactide-coglycolide (PLG) microspheres. It is packaged in vials as a sterile, white to off-white preservative-free, free-flowing powder.
Gel Product	Tube	Topical	Regranex® Gel, Becaplermin, Platelet-Derived Growth Factor
			Regranex gel is a non-sterile, low bioburden, preserved, sodium carboxy-methylcellulose-based topical gel, supplied in multi-use tubes.

4. ADEQUATE DESCRIPTION OF THE MANUFACTURING PROCESS

The FDA has indicated that they expect the manufacturer to make the following statement in their Phase 1 IND submission: "the sponsor believes that the manufacturing of either the drug substance or the drug product does not present any signal of potential human risk."[16] To support such a statement, the manufacturer needs to describe the drug product manufacturing process for the biopharmaceutical. This description needs to be accurate and of sufficient depth to allow the FDA reviewer to conclude that the manufacturing process is indeed safe and does not introduce any safety risks to the biopharmaceutical product to be administered to the patient.

The amount of information about the production process that the regulatory agencies expect to see included in the regulatory submissions increases as the clinical development advances. As stated in the FDA regulations, 21 CFR 312.23(a)(7)(i):

> "Although in each phase of the investigation sufficient information is required to be submitted to assure the proper identification, quality, purity and strength of the investigational drug, the amount of information needed to make that assurance will vary with the phase of the investigation, the dosage form, and the amount of information otherwise available."[34]

Keep in mind, that for biopharmaceuticals, the regulatory expectation is that "more information may be needed to assess the safety of biotechnology-derived drugs" compared to chemically-synthesized drugs.[16]

4.1. During Clinical Development

The CMC continuum applies to drug product manufacturing processes. At each stage of clinical development, the regulatory agencies expect that the minimum CMC requirements will be met, and that these requirements will increase and tighten as clinical development progresses.

However, when it comes to product safety, there is no continuum. The expectation is that adequate controls to prepare a safe product will be in place from the beginning of clinical trial drug product manufacturing. After all, should a patient on clinical study be put at risk any more than a patient who buys the drug product on the market?

4.1.1. Phase 1 IND Submission

For the Phase 1 IND, the amount of information needed to describe the drug product manufacturing process is expected to be 'brief':

> "A list of all components, which may include reasonable alternatives for inactive compounds, used in the manufacture of the investigational drug product, including both those components intended to appear in the drug product and those which may not appear, but which are used in the manufacturing process. A list of usually no more than one or two pages of written information should be submitted. The quality (e.g., NF, ACS) of the inactive ingredients should be cited. For novel excipients, additional manufacturing information may be required.
>
> A brief summary of the composition of the investigational new drug product should be submitted. In most cases, information on component ranges is not necessary.

The full street address(es) of the manufacturer(s) of the clinical trial drug product should be submitted.

A diagrammatic presentation and a brief written description of the manufacturing process should be submitted, including sterilization process for sterile products. Flow diagrams are suggested as the usual, most effective, presentations of this information."[16]

But what is 'brief'? My recommendation is that sufficient information about the drug product manufacturing process needs to be provided, especially about the aseptic controls, to assess safety of the biopharmaceutical. Ask yourself, what would you need to know about the drug product manufacturing process to make this assessment? That's what should be provided. Clearly, this doesn't mean submitting the batch production records. But it does mean summarizing the specifics of those batch production records and including them in the submission.

4.1.2. Phase 2 IND Submission

For Phase 2 submissions, since knowledge about the drug product manufacturing process and how to more adequately control it is expected to occur with time, additional description is expected to be filed, along with any changes that were made in the manufacturing process:

"Any changes to the information specified for Phase 1 (i.e., table listing of all components) should be provided. The components should be identified by their established names and compendial status, if they exist. In addition, quantitative composition per unit of use should be provided. A batch formula should be provided, if not already submitted.

For excipients, the quality (e.g., USP, NF) of the excipients should be specified if changed. Analytical procedures and acceptance criteria should be provided for noncompendial components. A brief description of the manufacture and control of these compounds or an appropriate reference should be provided (e.g., DMF, NDA, BLA). Information for excipients not included in previously approved drug products should be equivalent to that submitted for new drug substances.

Manufacturer: Updates on the information specified for Phase 1 should be provided.

A brief, step-by-step description of the manufacturing procedure for the unit dose should be provided. The description should focus only on the general manufacturing task. Flow diagrams should be included. Information does not need to be provided for the following: (1) the specific equipment used; (2) the packaging and labeling process; and (3) in-process controls, except for sterile products or atypical dosage forms. Only safety related information need to be submitted for reprocessing procedures and controls.

For sterile products, safety updates on the manufacturing process information filed for Phase 1 studies should be submitted. The Phase 2 information should include changes in the drug product sterilization process or other changes introduced in the process to sterilize bulk drug substance or bulk drug product, components, packaging, and related items. Information related to the validation of the sterilization process need not be submitted at this time.

A brief description of any changes in the container closure system (also referred to as packaging system) should be provided. The container closure system is defined as the sum of packaging components that together contain and protect the drug product."[36]

4.1.3. Phase 3 IND Submission

By the time of the Phase 3 submissions, a much greater amount of drug product manufacturing description is expected to be filed:

"The sponsor should provide updated information regarding the components, composition and batch formula for Phases 1 and 2. Components that are removed during the manufacturing of the drug product should be listed, but quantitative values do not need to be reported. Quantitative information should be reported for the batch formula.

A listing of all firms associated with the manufacturing and controls of the drug product should be submitted, including the contractors for stability studies, packaging, labeling and quality control release testing.

A general step-by-step description of the manufacturing method for a unit dose should be provided, including key equipment employed. Where the qualitative formulation does not change, a single description of the manufacture of different strength unit doses can be used. The description should indicate how the material is being processed and can be general enough to allow for flexibility in development. In planning the clinical batch size, the sponsor should consider the postapproval production scale. A brief description of the packaging and labeling process for clinical supplies should be provided. Reprocessing procedures and pertinent controls should be described, if applicable.

For sterile products, updates on information specified for Phases 1 and 2 should be provided. The information should include a description of changes in the drug product sterilization process or other changes introduced into the process to sterilize bulk drug substance or bulk drug product, components, packaging, and related forms. Information related to the validation of the sterilization process need not be submitted at this time, but should be submitted at the time of an NDA or BLA filing.

Container Closure System: Updates on the information previously filed should be submitted. In addition, the name of the manufacturer and supplier should be provided. If the component meets USP criteria, it should be stated (e.g., Type 1 glass). A DMF reference and authorization should be provided, if available.

Labeling: Updates on the information filed for Phase 1 should be provided during Phases 2 and 3."[36]

4.2. BLA/NDA Submission

The 'complete' description of the drug product manufacturing process is required in the market approval submission. And when the FDA says 'complete', they mean it. In fact, if this manufacturing information is not complete, it can be a cause for the FDA to refusal to file the submission.[31] Some of the CMC information on the drug product manufacturing process is presented in Table 56.

Table 56. CMC information to be included in the BLA/NDA dossier to describe the drug product manufacturing process for recombinant proteins and monoclonal antibodies[40]

Composition	A tabulated list of all components with their unit dose and batch quantities for the drug product or diluent should be submitted.
	The composition of all ancillary products that might be included in the final product should be included.
Specifications and Methods for Drug Product Ingredients	The specifications for all ancillary products that are included in the product should be provided.
	Information on all excipients including process gases and water should be included. A list of compendial excipients and the citations for each should be submitted. For non-compendial excipients, tests and specifications should be described. For a novel excipient, the description should include its preparation, characterization, and controls. For inactive ingredients of human or animal origin, certification, results of testing or other procedures, or validation data demonstrating their freedom from adventitious agents should be provided.
Manufacturer (s)	The name(s) and address(s) of all manufacturers involved in the manufacture and testing of the drug product including contractors, and a description of the responsibility(ies) of each should be submitted.
	A list of other products (R&D, clinical or approved) made in the same rooms should be provided.

Methods of Manufacturing and Packaging	A complete description of the manufacturing process flow of the formulated bulk and finished drug product should be provided. This discussion should include a description of sterilization operations, aseptic processing procedures, lyophilization, and packaging procedures.
	Accompanying this narrative, a flow chart should be provided that indicates the production step, the equipment and materials used, the room or area where the operation is performed and a listing of the in-process controls and tests performed on the product at each step. This flow diagram or narrative should also include information on the methods of transfer of the product between steps (i.e., sterile, SIP connection, sanitary connection, open transfers under laminar flow units, etc. Such transfers should be described for movement of product between equipment, areas/rooms, buildings and sites.
Container/Closure System(s)	A description of the container and closure system, and its compatibility with the drug product should be submitted. Detailed information concerning the supplier(s), address(es), and the results of compatibility, toxicity and biological tests should be included. Alternatively, a DMF can be referenced for this information.
	For sterile product, evidence of container and closure integrity should be provided for the duration of the proposed expiry period.
Microbiology	Information should be submitted as described in the 'Guidance for Industry for the Submission of Documentation for Sterilization Process Validation in Applications for Human and Veterinary Drug Products'.

The FDA has prepared a guidance document describing how to organize the required drug product manufacturing CMC content into the Common Technical Document (CTD) format.[42]

5. ADEQUATE CONTROL OVER THE MANUFACTURING PROCESS

Most biopharmaceuticals since they are injectables are required to be 'sterile'. There is a significant public health implication of distributing a non-sterile drug purporting to be sterile. The FDA has reported that nearly all drugs recalled due to 'non-sterility' or 'lack of sterility assurance' in the period spanning 1998 to 2000 were produced via aseptic processing.[121] There should be no surprise at the significant attention that the FDA is placing on validation of these processes.

As with all other manufacturing operations, the regulatory requirement is that the drug product manufacturing process steps need to be defined, controlled, monitored and validated.

5.1. Regulatory Requirements for Market Approved Products

Since the regulatory requirements for a drug product manufacturing process apply equally to both chemically-synthesized drug products and biopharmaceutical drug products (e.g., cGMP operations, validation of the equipment and the process, protection from contamination and cross-contamination, label controls, etc.), these requirements will not be covered in great detail in this book. However, the issue of aseptic processing is central to biopharmaceutical drug products so it will be discussed.

The FDA has provided extensive guidance on what they expect in order for a manufacturer to safely produce sterile drugs by aseptic processing:

- Guideline on Sterile Drug Products Produced by Aseptic Processing (1987)[122]

- Guidance for Industry for the Submission of Documentation for Sterilization Process Validation in Applications for Human and Veterinary Drug Products (1994)[123]

- Sterile Drug Products Produced by Aseptic Processing (Draft, 2002)[121]

The European GMPs have a similar guidance for aseptic processing:

- Manufacture of Sterile Medicinal Products (2003)[93]

In a published paper, FDA reviewers elaborated on their expectations for the environmental conditions for the drug product manufacturing of biopharmaceuticals under aseptic processing conditions:

"Formulation and final product preparation are considered critical areas when product is exposed to the environment and not exposed to any subsequent downstream processing. The areas should be Class 100 and should meet the action limits for viable and nonviable particulates. Areas immediately surrounding Class 100 filling areas should be at least Class 10,000. Formulation of final bulk subject to subsequent sterilization should occur in a controlled environment with an air classification of at least 100,000. We recommend that open steps be performed in a Class 100 environment, such as a laminar flow hood, within the controlled area.

Monitoring frequencies for viable and nonviable particulates in critical areas should occur daily during operations. Viable particulate monitoring should consist of active volumetric air sampling and surface sampling using contract plates and/or swabs for all equipment, work surfaces, floors, walls, and personnel. Active volumetric air sampling should occur in close proximity to operations and in areas of heavy personnel traffic. Alternative sampling methods, such as settling plates, may be incorporated into the environmental monitoring protocol to provide additional information for correlation with active volumetric air sampling.

The area surrounding the filling area should be, at minimum, designated as Class 10,000 under dynamic conditions in order to ensure that appropriate conditions are maintained within the encompassed Class 100 filling area. These areas should also be monitored during each fill."[91]

For clarification, clean room classifications are based on limits for nonviable and viable airborne particulate monitoring, microbial surface monitoring, and personnel monitoring; Class 100 is also referred to as Grade A or Class M3.5 or ISO14644-1 5; Class 10,000 is also referred to as Class C or M5.5 or ISO14644-1 7; Class 100,000 is also referred to as Grade D or Class M6.5 or ISO14644-1 8.[92, 93]

There are many possible sources of contamination of the cleanroom environment. Equipment, structures, and surfaces can generate particles through friction, heat, exhaust, outgassing, and static electricity. Incoming production components (especially cardboard) also introduce contaminants. But it is the personnel that work in the aseptic processing environment that generate the most particles. As manufacturing operators carry out their operations and quality personnel collect environmental samples, they generate millions of particles with every movement (Table 57).

Table 57. Generation of particles by people working in a cleanroom[124]

Activity in Cleanroom	Particle Generation Rate ($\geq$ 0.3 microns per minute)
Motionless/ sitting/ standing	100,000
Standing to sitting position	2,500,000
Walking at 2.0 mph	5,000,000
Walking at 3.5 mph	7,500,000
Walking at 5.0 mph	10,000,000

Barrier isolator technology is one answer to controlling the personnel generation of particles during aseptic processing. A barrier isolator filling system encloses the entire filling area to prevent microbial and particulate contamination prior to and during filling, check weighing and stoppering operations.

5.2. Regulatory Expectations During Clinical Development

As discussed in Chapter 4, cGMPs, the regulatory agencies expect the drug product manufacturing process to yield a safe and sterile (if it is to be injected into the patient) drug product, even at Phase 1. However, the FDA recognizes that during clinical development, a company will be developing their drug product manufacturing control systems, improving their documentation, completing their test method validation, etc.

Europe on the other hand has released a proposed set of requirements for medicinal preparations to be used in clinical studies prepared by aseptic processing (Table 58)

Table 58 European regulatory expectations during clinical development for the manufacture of the drug product[49]

General Principles	The application of GMP to the manufacture of investigational medicinal products is intended to ensure that trial subjects are not placed at risk, and that the results of clinical trials are unaffected by, inadequate safety, quality or efficacy arising from unsatisfactory manufacture. Equally it is intended to ensure that there is consistency between batches of the same investigational medicinal product used in the same or different clinical trials or that changes during the development of an investigational medicinal product are adequately documented and justified.
	The production of investigational medicinal products involves added complexity in comparison to marketed products by virtue of the lack of fixed routines, variety of clinical trial designs, consequent packaging designs, the need, often, for randomisation and blinding and increased risk of product cross-contamination and mix up. Furthermore there may be incomplete knowledge of the potency and toxicity of the product and a lack of full process validation.
	These challenges require personnel with a thorough understanding of, and training in, the application of GMP to investigational medicinal products.
	This increased complexity in manufacturing operations requires a highly effective Quality system.
Sterile Drug Products	For sterile products, the validation of sterilizing processes should be of the same standard as for products authorized for marketing
Validation of Aseptic Processing	Validation of aseptic processes presents special problems when the batch size is small; in these cases the number of units filled may be the maximum number filled in production. If practicable, and other wise consistent with simulating the process, a larger number of units should be filled with media to provide greater confidence in the results obtained. Filling and sealing is often a manual or semi-automated operation presenting great challenges to sterility so enhanced attention should be given to operator training, and validating the aseptic technique of individual operators.
	Where processes such as mixing have not been validated, additional quality control testing may be necessary.

Control of the Manufacturing Process	During the development phase, validated procedures may not always be available, which makes it difficult to know in advance the critical parameters and the in-process controls that would help to control these parameters. In these cases, provisional production parameters and in-process controls may be deduced from experience with analogues and from earlier development work.
	Careful consideration by key personnel is called for in order to formulate the necessary instructions and to adapt them continually to the experience gained in production. Parameters identified and controlled should be justifiable on knowledge available at the time.
	As processes may not be standardized or fully validated, testing takes on more importance in ensuring that each batch meets its specification.
Cross-Contamination Controls	The toxicity, potency and sensitizing potential may not be understood for investigational medicinal products and this reinforces the need to minimize the risks of cross-contamination.
	Cleaning of premises and equipment between different products is of particular importance. Cleaning should be very stringent with procedures designed in the light of the incomplete knowledge of the toxicity of the investigation medicinal product.

When it comes to safety, patients on clinical trial materials deserve as much drug product safety as that expected by drug products purchased on the market!

6. WHAT CAN GO WRONG

With all of the things that need to be properly controlled during drug product manufacturing of a biopharmaceutical, companies in their haste can fail to meet the requirements expected by the regulatory agencies. This usually results in inspectional issues when the FDA arrives at the manufacturing facility. Table 59 presents some drug product manufacturing control deficiencies recorded by the FDA.

Table 59. Some drug product manufacturing deficiencies recorded by FDA inspectors[50, 125]

<u>Warning Letter to Centocor July 10, 1998</u>

During sterile media fill operations, not all glass vials that are filled with growth media are incubated to detect microbiological growth. For example, vials with the incorrect volume of growth media, vials with incorrectly seated rubber stoppers, and vials with defective metal crimps are discarded.

Sterile media fill vials are not incubated at an optimum temperature to promote mold growth, despite the identification of mold contaminates recovered from manufacturing areas during routine environmental monitoring.

There are no written procedures to describe the placement of thermocouples and biological indicators used during the sterilization validation runs for the autoclaves.

Immediately over the crimping station there was a light fixture panel that contained dirt, dust, and a water stain on the interior surface.

<u>Warning Letter to Genentech, Inc., December 14, 2000</u>

Failure to obtain approval from the quality control unit prior to reprocessing in that, there is no documentation that the quality control unit was notified prior to the re-filtration of Activase lot #L9042A. On April 10, 2000, during set up for filling, manufacturing detected a leak in the connection during priming. On April 11, 2000, the bulk was re-filtered by manufacturing and filled.

On January 15, 2000, during the manufacture of Pulmozyme bulk lot, an expired concentrated bulk lot was used. There is no indication that the impact to material, which was held for an extended period of time, was evaluated.

Blue-gray particulates were observed in Pulmozyme thawed bulk lot. The bulk was released for filtration. Final vial lot was released by Quality Assurance. However, there is no approved SOP to allow for re-filtration due to particles and the incident was not reported to the agency.

---

7. STRATEGIC CMC TIPS FOR DRUG PRODUCT MANUFACTURING

Most of the aseptic processing controls need to be in place for the manufacture of Phase 1 clinical supplies. But for an aseptic processing operation, there is a difference between clinical operations and commercial operations, especially due to all of the attention that aseptic processing is receiving from the FDA. Therefore, it may be wise to consider moving the drug product manufacturing process for later stage clinical supplies to a facility that already has experience in working with market approved products. In this way, further assurance can be obtained that the biopharmaceutical drug product manufacturing process will pass the pre-approval inspection.

Formulation development most likely will be very limited at Phase 1 due to the race to initiate the clinical studies. However, during the course of clinical development, studies should be initiated to evaluate the formulation, container-closure and process attributes that can influence batch reproducibility, product performance, and drug product quality. By the time of BLA/NDA filing, justification of the pharmaceutical development is required:

> "The Pharmaceutical Development section should contain information on the development studies conducted to establish that the dosage form, formulation, manufacturing process, container closure system, microbiological attributes, and usage instructions are appropriate for the purpose specified in the application."[42]

Figure 13 presents some aspects of the minimum CMC continuum for a biopharmaceutical drug product manufacturing process.

Figure 13. Minimum CMC continuum for a biopharmaceutical drug product manufacturing process

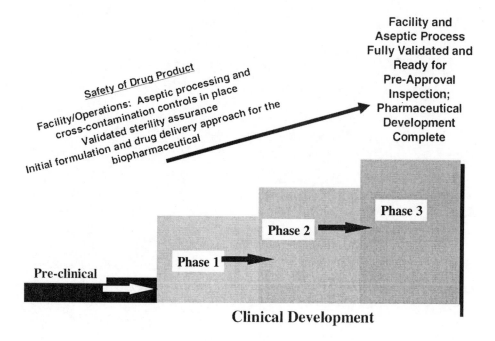

9

Physicochemical/Biological Analysis of the Biopharmaceutical Product

'You are not currently required to submit samples of future lots of product to the Center for Biologics Evaluation and Research (CBER) for release by the Director, CBER, under 21 CFR 610.2. FDA will continue to monitor compliance with 21 CFR 610.1 requiring assay and release of only those lots that meet release specifications.

The dating period for this product shall be 12 months from the date of manufacture when stored at 2-8°C. The date of manufacture shall be defined as the date of final sterile filtration of the final formulated product. The drug substance may be stored for up to 12 months at 2-8°C.

Results of ongoing stability studies should be submitted throughout the dating period as they become available including the results of stability studies from the first three production lots. The stability protocol in your license application is considered approved for the purpose of extending the expiration dating period of your drug substance and drug product as specified in 21 CFR 601.12.'

Typical wording found in a CBER product approval letter

1. A CHALLENGING ANALYSIS

Biopharmaceuticals are complex products that are either composed of amino acids with or without carbohydrates (these are the recombinant proteins and the monoclonal antibodies) or composed of nucleic acids (these are the plasmid gene vectors and the viral vectors used for gene transfer). Especially for the proteins, their analysis can be a major challenge, since even a minor primary structural change or conformational change could impact the therapeutic activity and/or the immunogenicity of the molecule.

1.1. Goals: Consistent, Safe, Potent and Pure Product

To achieve the regulatory requirement of demonstrating that the product is consistently manufactured, safe, potent and pure, the biopharmaceutical (1) must be thoroughly characterized; (2) must meet all of its defined, scientifically sound and appropriate specifications; and (3) must continue to meet all of the defined specifications throughout its entire shelf-life.

This goal is accomplished by both characterization of the biopharmaceutical and testing during release and on stability:

- 'Product characterization' is precisely deciphering and describing a product's physicochemical and biological properties, using the plethora of methods available to examine these biopharmaceuticals. Product characterization is performed to learn more about the science of the biopharmaceutical molecules, to understand where they may be chemically weak and unstable, and to try to understand what changes may occur that could impact clinical efficacy or safety.

- 'Product release testing' is that testing, performed by Quality Control (QC), in order to release a specific lot of product. Based on the knowledge obtained by product characterization, a subset of scientifically sound and appropriate release tests is chosen. For each product release test, there is an assigned 'specification'. A specification is defined as "a list of tests, references to analytical procedures, and appropriate acceptance criteria which are numerical limits, ranges, or other criteria for the tests described."[126] The specification establishes the set of criteria to which a product must conform to in order to be considered acceptable for its intended use.

- 'Product stability testing' is that testing, performed by QC, in order to assure that the released product continues to meet specifications. From the subset of release tests, those that are stability-indicating are used for stability testing of the product. 'Product shelf life' is the period of time, at defined storage conditions, that the product is expected to remain within the specification limits. The 'expiration date' is "the calendar month and year, and where applicable the day and hour, that the dating period ends."[127]

1.2. Relationship Between Product Characterization and QC Testing

The relationship between product characterization and the QC release and stability testing is illustrated in Figure 14.

Each type of testing has its purpose and this purpose shouldn't get confused. Unless product characterization testing has been thorough, the selection of QC release tests may not be appropriate or even meaningful. There is no value in running every test method available in the stability program, only those test methods that have been demonstrated to be stability-indicating.

Figure 14. Testing pyramid illustrating the relationship between product characterization and QC release and stability testing

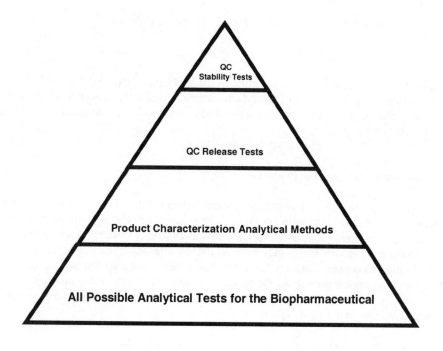

2. UNRAVELING THE MOLECULAR PROPERTIES

Each biopharmaceutical is defined by its own set of molecular properties. The general molecular properties include a given molecular weight and primary structure. Each biopharmaceutical also exists in higher order structures (i.e., conformation for proteins and coiling for DNAs). But the challenge in analyzing biopharmaceuticals is in detecting the abundance of multiple molecular forms, referred to as molecular variants, which occur in the biologically-expressed products.

2.1. Molecular Variants for DNA

A few molecular variants occur with the DNA-based biopharmaceuticals, and they can arise from the following three major sources:

- Genetic mutation leading to changes in nucleic acid bases

- Recombination of virus vectors leading to replacement of nucleic acids for replication competence

- Supercoiling versus linear forms of plasmid DNA

For the DNA-based biopharmaceuticals (the plasmid DNA vectors and the viral vectors), there is a limited set of analytical methods available. These methods include DNA sequencing and gel electrophoresis (with and without site-specific restriction enzyme treatment).

2.2. Molecular Variants for Proteins

However, for the recombinant proteins and monoclonal antibodies, an abundance of molecular variants is frequently observed when using today's sophisticated analytical methodology, commonly referred to as product heterogeneity or microheterogeneity.
These multiple molecular forms may have no measurable impact on the therapeutic effect of the biopharmaceutical, but on the other hand they may also impact the biological activity and/or immunogenicity of the product. Predicting this impact is difficult.
For the recombinant proteins and monoclonal antibodies, the molecular variants can be formed by four major pathways:

- Precursors of the desired biopharmaceutical product

- Post-translational modifications

- Degradation products arising during manufacture

- Degradation products arising during storage and shipment

Table 60 presents some of the many molecular variants which can occur with these biopharmaceuticals.

Table 60. Some molecular variants of recombinant proteins and monoclonal antibodies

Missing Terminal Amino Acids	N-terminal and/or C-terminal post-translational amino acid processing
Substituted Amino Acids	Amino acids on the protein that do not match the predicted DNA sequence (e.g., substitution of norleucine for methionine)
Glycosylation	Carbohydrate residues attached post-translationally to protein as it is released from the ribosome; attached either to serine or threonine ('O-linked') or to asparagine ('N-linked')
Chemical Modification of Amino Acids	Oxidation methonine → methonine sulfoxide
	Deamidation: asparagine → aspartic acid
	N-terminal cyclization: glutamine → pyroglutamic acid
Fragmentation	Breaking of the protein chain either by chemical hydrolysis or by trace proteolytic enzyme impurity
Disulfide Exchange	Intramolecular disulfide exchange resulting in misfolded conformation
	Intermolecular cross linking resulting in oligomerization (dimers, trimers, etc.)
Aggregation	Multiple associated proteins held together through hydrophobic forces
	Multiple protein molecules chemically linked through amide bonding (de-hydrolysis of amino acids on different protein molecules)

Glycosylation is frequently the major source of observed microheterogeneity in biopharmaceuticals.

2.3. Plethora of Analytical Methods Available For Proteins

It is not surprising that the level of detail and complexity typifying modern analytical techniques parallels that of the biopharmaceutical products themselves. The suite of techniques used includes chromatographic, electrophoretic and spectroscopic methods, each further comprising a host of techniques. Brief descriptions of some of the methods available today are described below:

- Amino Acid Analysis

    Complete hydrolysis of the protein followed by chromatographic separation of the amino acids present; typically used to determine the amino acid composition of the product

- Amino Acid Sequencing

    Edman Sequencing: Selective chemical removal of amino acids one at a time from the N-terminal end of the protein; typically used to sequence the first 15-20 amino acids into the protein

    Carboxy-Terminal Sequencing: Enzymatic or chemical step wise degradation of the protein molecule from the C-terminus followed by identification of the specific amino acid(s) released

- Mass Spectrometry (MS)

    Molecules are ionized and then the ions are separated by mass/ionic charge ratio; the ionic charge of most molecules is 1 so separation is by mass

- Chromatographic Profiling by High Performance Liquid Chromatography (HPLC)

    Size Exclusion: resolves proteins by their hydrodynamic radius of gyration (i.e., molecular size compared to a globular molecule)

    Ion Exchange: resolves proteins by their charge

    Reversed Phase: resolves proteins by their hydrophobicity

    Hydrophobic Interaction: resolves proteins by their hydrophobicity

- Peptide Mapping

    The protein is first digested with site-specific proteases (e.g., trypsin to cleave proteins at lysine and arginine sites, Lys C to cleave after lysine sites, etc.) and then the peptides formed are resolved by HPLC;

frequently coupled with MS to obtain a complete sequence analysis of the protein

- Electrophoretic Profiles by Molecular Size

  SDS Polyacrylamide Gel Electrophoresis (SDS-PAGE): Proteins are unfolded and complexed with the ionic detergent sodium dodecyl sulfate (SDS) to give a uniform charge to all of the molecular species; the proteins are then sieved through the gel matrix in an electric field which resolves proteins by molecular size; resolved proteins are then visualized either by a colorimetric stain (e.g., Coomassie Blue, silver staining) or immunoblotting (e.g., Western blots with specific labelled antibodies); performed under reducing and/or non-reducing conditions

  Capillary Electrophoresis-SDS (CE-SDS): Similar in result to SDS-PAGE but of a higher resolution; use of a tiny capillary column rather than a flat gel to achieve the sieving

- Electrophoretic Profiles by Molecular Charge

  Isoelectric Focusing (IEF): Molecular charge variants are resolved in a charged pH gradient gel; protein molecules migrate to their pI value (i.e., the net molecular charge is 0) when an electric field is applied; resolved proteins are then visualized by a colorimetric stain (e.g., Coomassie Blue, silver staining)

  Capillary Isoelectric Focusing (cIEF): Similar in result to IEF but of a higher resolution; use of a tiny capillary column rather than a flat gel

- Dynamic Light Scattering

  Time-dependent fluctuations in scattered intensity caused by Brownian motion and diffusion of molecules related to the molecular size of a globular protein

- Circular Dichroism (CD)

  Measures the absorbance difference between right-handed and left-handed circularly polarized light as a function of wavelength; sensitive to symmetry around the amino acid chromophores

- Ultracentrifugation

  Sedimentation velocity relates directly to molecular weight and conformation

Each analytical method has its strength and limitations when it comes to the analysis of the recombinant proteins and monoclonal antibodies. This is illustrated in Table 61. It should also be noted that the analytical test methods typically provide no useful information concerning the biological activity of the product.

Table 61. Data possible from the plethora of analytical methods available to evaluate recombinant proteins and monoclonal antibodies

Test Method	Size	Charge	1° Structure	Aggregation	Purity	Potency
Mass Spectrometry (MS)	+++	+	+++	-	++	-
Circular Dichroism (CD)	-	-	-	+++	+	-
HPLC: Size Exclusion	+++	-	-	+++	++	-
HPLC: Ion Exchange	-	+++	++	+	++	-
HPLC: Reverse Phase	+	+	+++	-	+++	-
SDS-PAGE	+++	-	++	-	+++	-
Isoelectric Focusing (IEF)	-	+++	++	+	++	-

Number of +'s indicate usefulness of test method in determining a specific property

The full understanding of a biopharmaceutical comes only when a full range of analytical methods are applied. Obtaining this full understanding is the goal of product characterization.

3. CHARACTERIZATION OF BIOPHARMACEUTICALS

The complexity of the structure and function of biopharmaceuticals necessitates a wide range of analytical procedures to adequately characterize the product. Since no single test method is a stand-alone solution to a full understanding of all of the possible changes that can occur with the molecule, and since no single test method provides the same depth of useful information across all biopharmaceutical products, many analytical and biological methods should be performed at least once to determine whether useful information about the molecule can be obtained.

While the science uncovered during characterization will be interesting, the value of the characterization will be three-fold:

1. To guide the selection of relevant release assays and meaningful specifications

2. To guide the selection of relevant stability-indicating assays and meaningful expiration dates

3. To develop analytical and biological tools to assess the impact of any future process changes

Characterization is frequently considered a single event. However, it will need to be repeated through clinical development whenever significant process changes are implemented.[126]

3.1. Regulatory Expectations During Clinical Development

The minimum CMC continuum applies to characterization of a biopharmaceutical. At each stage of clinical development, the regulatory agencies expect that the minimum characterization requirements will be met, and that these requirements will increase and tighten as clinical development progresses. This is indicated in the FDA's guidance documents on what they expect to see included in the IND submissions.

3.1.1. Phase 1 IND Submission

"A brief description of the drug substance and some evidence to support its proposed structure should be submitted. It is understood that the amount of structure information will be limited in the early stage of drug development."[16]

The FDA recognizes that the amount of characterization will be limited in the early stage of drug development.   However, for a biopharmaceutical, much more characterization is expected in the early clinical stages:

> "Before a mAb is studied in humans, a precise and thorough characterization of antibody structural integrity, specificity, and potency should be conducted and described in the IND."[29]

3.1.2. Phase 2 IND Submission

> "Safety updates on information identified in the phase 1 guidance (i.e., a brief description of the drug substance and some evidence to support its proposed chemical structure) should be provided, with a more detailed description of the configuration and chemical structure for complex organic compounds.

> The sponsor can select a batch to be used as a reference material, against which initial clinical batches are tested prior to their release. Preferably, the sponsor should establish a working standard even at the initial stage of development. Where a recognized national or international standard (primary standard) is available, the manufacturer's reference material and/or working standard should be calibrated against this standard." [36]

Reference materials are expected to be in use during Phase 2.

3.1.3. Phase 3 IND Submission

> "A complete description of the physical, chemical, and biological characteristics of the drug substance should be provided including elements such as ... isoelectric point (pI) ... Ig class for immunoglobulins and biological activities (when applicable).

> For peptides and proteins, characterization should include data on the amino acid sequence, peptide map, post-translational modifications (e.g., glycosylation, gamma carboxylation), and secondary and tertiary structure information, if known.

> If a national or international standard is not yet available, the sponsor should establish its own primary reference material during phase 3 studies. The manufacturer can continue to use the working standard used in phase 2 or can establish a new working standard for lot release. The synthesis and purification of the reference material or working standard used should be described if it differs from that of the investigational drug substance. The analytical

procedures and calibration result for the working standard against the primary reference material should be provided." [36]

By Phase 3, the expectation is that further characterization information is available.

3.2 Regulatory Expectations for the BLA/NDA Submission

By the time of filing of the dossier for market approval, characterization of the biopharmaceutical is expected to be complete and thorough. Table 62 presents the characterization information that is expected to be included in the filing.

Table 62. Characterization information to be included in the API section of the BLA/NDA for recombinant proteins and monoclonal antibodies[40, 41]

Characterization	A description and the physicochemical results of all the analytical testing performed on the manufacturer's reference standard lot and qualifying lots to characterize the drug substance should be included.
	Information from specific tests regarding identity, purity, stability and consistency of manufacture of the drug substance should be provided.
	Additional physicochemical characterization may be required for products undergoing post-translational modifications, for example, glycosylation, sulfation, phosphorylation, or formulation.
	All test results should be fully described and the results provided. The application should also include the actual data such as legible copies of chromatograms, photographs of SDS-PAGE or agarose gel, spectra, etc.
	A description and results of all relevant in vivo and in vitro biological testing performed on the manufacturer's reference standard lot to show the potency and activities of the drug substance should be provided. Results of other relevant testing performed on lots other than the reference standard lot, that might have been used in establishing the biological activity of the product, should also be included.
	The description and validation of the bioassays should include the methods and standards used, the inter- and intra-assay variability, and acceptable limits of the assay.
Impurities Profile	A discussion of the impurities profiles, with supporting analytical data, should be provided.
	Profiles of variants of the protein drug (e.g., cleaved, aggregated, deamidated, oxidized forms, etc.), as well as non-product related impurities (e.g., process reagents and cell culture components), should be included.
Reference Standard	If an international reference standard (WHO, NIBSC) or compendial reference standard (USP) is used, submit the citation for the standard and a certificate of analysis.

If no biological potency or chemical reference standard exists, and the applicants establish their own primary reference standard, a description of the characterization of the standard should be provided. Submit the results of testing, such as physicochemical and biologic activity determinations, of the standard and provide a certificate of analysis. The Standard Operating Procedures (SOPs) to be used for

qualifying a new reference standard should be provided. Information should also be provided on the stability of any reference standard.

If an in-house working reference standard is used, a description of the preparation, characterization, specifications, testing and results should be provided. The data from the calibration of the in-house working reference standards against a primary reference standard should also be submitted.

The characterization must be complete in the filing. A specified reference standard lot needs to be identified, in addition to a procedure to qualify future reference standards. Along with a thorough grasp of the impurity profile for the biopharmaceutical, there is the need to have a strong handle on the biological activity of the product.

3.3. Full Characterization of Recombinant Proteins and Monoclonal Antibodies

The full characterization of a recombinant protein or a monoclonal antibody requires considerable resource and time. Since so many of these biopharmaceuticals have been reviewed for market approval by the FDA, they have considerable experience in understanding what is important. These products have considerable molecular variants that need to be detected and studies. Special issues of host-dependent glycosylation and production-dependent impurities have to be addressed.

3.3.1. What is 'Full' Characterization?

The ICH Q6B guidance document on 'Test Procedures and Acceptance Criteria for Biotechnological/Biological Products'[126] provides insight into the amount of effort required to fully characterize a recombinant protein or monoclonal antibody (Table 63).

Table 63. Analyses to consider for full characterization of a recombinant protein and monoclonal antibody[126]

Amino Acid Composition	The overall amino acid composition is determined using various hydrolytic and analytical procedures, and compared with the amino acid composition deduced from the gene sequence for the desired product, or the natural counterpart.
	In many cases, amino acid composition analysis provides some useful structural information for peptides and small proteins, but such data are generally less definitive for large proteins.
	Quantitative amino acid analysis data can also be used to determine protein content in many cases
Terminal Amino Acid Sequence	Terminal amino acid analysis is performed to identify the nature and homogeneity of the amino- and the carboxy-terminal amino acids. If the desired product is found to be heterogeneous with respect to the terminal amino acids, the relative amounts of the variant forms should be determined using an appropriate analytical procedure. The sequence of these terminal amino acids should be compared with the terminal amino acid sequence deduced from the gene sequence of the desired product.
Peptide Map	Selective fragmentation of the product into discrete peptides is performed using suitable enzymes or chemicals and the resulting peptide fragments are analyzed by HPLC or other appropriate analytical procedure. The peptide fragments should be identified to the extent possible using techniques such as amino acid compositional analysis, N-terminal sequencing, or mass spectrometry.
	Peptide mapping of the drug substance or drug product using an appropriately validated procedure is a method that is frequently used to confirm desired product structure for lot release purposes.
Sulfhydryl Groups and Disulfide Bridges	If based on the gene sequence for the desired product, cysteine residues are expected, the number and positions of any free sulfhydryl groups and/or disulfide bridges should be determined to the extent possible. Peptide mapping (under reducing and non-reducing conditions), mass spectrometry, or other appropriate techniques may be useful for this evaluation.
Carbohydrate Structure	For glycoproteins, the carbohydrate content (neutral sugars, amino sugars, and sialic acids) is determined. In addition, the structure of the carbohydrate chains, the oligosaccharide pattern (antennary profile) and the glycosylation site(s) of the polypeptide chain is analyzed, to the extent possible.
Molecular Weight or Size	Molecular weight (or size) is determined using size exclusion chromatography, SDS-polyacrylamide gel electrophoresis (under reducing and non-reducing conditions), mass spectrometry, and other appropriate techniques.
Isoform Pattern (Charge)	This is determined by isoelectric focusing or other appropriate techniques.
Extinction Coefficient (Molar Absorptivity)	In many cases it will be desirable to determine the extinction coefficient (or molar absorptivity) for the desired product at a particular UV/visible wavelength (e.g., 280 nm). The extinction coefficient is determined using UV/visible spectrophotometry on a solution of the product having a known protein content as determined by techniques such as amino acid compositional analysis or nitrogen determination, etc.
	If UV absorption is used to measure protein content, the extinction coefficient for the particular product should be used.

Electrophoretic Patterns	Electrophoretic pattern and data on identity, homogeneity, and purity can be obtained by polyacrylamide gel electrophoresis, isoelectric focusing, SDS-polyacrylamide gel electrophoresis, Western-blot, capillary electrophoresis, or other suitable procedures.
Liquid Chromatographic Patterns	Chromatographic patterns and data on the identity, homogeneity, and purity can be obtained by size exclusion chromatography, reversed-phase liquid chromatography, ion-exchange liquid chromatography, affinity chromatography or other suitable procedures.
Spectroscopic Profiles	The ultraviolet and visible absorption spectra are determined as appropriate. The higher-order structure of the product is examined using procedures such as circular dichroism, nuclear magnetic resonance (NMR), or other suitable techniques, as appropriate.
Biological Activity	Assessment of the biological properties constitutes an equally essential step in establishing a complete characterization profile.
Cell Substrate-Derived Impurities	Cell substrate-derived impurities include, but are not limited to, proteins derived from the host organism, nucleic acid (host cell genomic, vector, or total DNA).
	For host cell proteins, a sensitive assay (e.g., immunoassay, capable of detecting a wide range of protein impurities) is generally utilized. In the case of an immunoassay, a polyclonal antibody used in the test is generated by immunization with a preparation of a production cell minus the product-coding gene, fusion partners, or other appropriate cell lines.
	The level of DNA from the host cells can be detected by direct analysis on the product (such as hybridization techniques).
	Clearance studies, which could include spiking substrate-derived impurities such as nucleic acids and host cell proteins may sometimes be used to eliminate the need for establishing acceptance criteria for these impurities.
Cell Culture-Derived Impurities	Cell culture-derived impurities include, but are not limited to, inducers, antibiotics, serum, and other medium components.
Downstream-Derived Impurities	Downstream-derived impurities include, but are not limited to, enzymes, chemical and biochemical processing reagents (e.g., cyanogens bromide, guanidine, oxidizing and reducing agents), inorganic salts (e.g., heavy metals, arsenic, non-metallic ion), solvents, carriers, ligands (e.g., monoclonal antibodies), and other leachables.
Product-Related Impurities	Truncated forms, modified forms, and aggregates are the most frequently encountered molecular variants of the desired product.
	Such variants may need considerable effort in isolation and characterization in order to identify the type of modification.
	Truncated forms: Hydrolytic enzymes or chemicals may catalyze the cleavage of peptide bonds. These may be detected by HPLC or SDS-PAGE. Peptide mapping may be useful, depending on the property of the variant.
	Other modified forms: Deamidated, isomerized, mismatched S-S linked, oxidized or altered conjugated forms (e.g., glycosylation, phosphorylation) may be detected and characterized by chromatographic, electrophoretic and/or other relevant analytical methods (e.g., HPLC, capillary electrophoresis, mass spectrometry, circular dichroism).

Aggregates: The category of aggregates includes dimmers and higher multiples of the desired product. These are generally resolved from the desired product and product-related substances, and quantitated by appropriate analytical procedures (e.g., size exclusion chromatography, capillary electrophoresis).

With so many monoclonal antibodies either approved for market by the FDA or in clinical study, the 'norm' of characterization for these products is quite extensive, as illustrated by the example of the monoclonal antibody palivizumab (Synagis®) in Table 64.

Table 64. Characterization reported for the humanized monoclonal antibody palivizumab (Synagis®)[63, 128]

Characterization Test	Structural Level	Use
Amino acid analysis	Primary	Verify amino acid content compared to that predicted by the cDNA
N-terminal sequencing	Primary	Verify the integrity of the N-terminus
In situ CNBR fingerprint sequencing	Primary	Verify the integrity of selected internal sequences
UV spectrum	Primary	Verify that spectral profile is consistent with a protein
IgG isotype	Primary/secondary/tertiary	Confirm IgG subclass
Peptide mapping with MALDI-TOF MS or LC-ESMS	Primary	Confirm primary structure and posttranslational modifications
MALDI-TOF mass spectrometry	Primary	Verify molecular weight of intact IgG, identify impurities by mass
Monosaccharide composition analysis	Primary	Determine carbohydrate content

Oligosaccharide profile analysis	Primary	Distinguish and identify oligosaccharide heterogeneity
SDS-PAGE (non-reducing)	Primary/secondary/tertiary	Determine purity and identify aggregates
SDS-PAGE (reducing)	Primary	Determine purity by detecting breakdown products
Capillary gel electrophoresis (CGE)	Primary/secondary/tertiary	Determine purity by detecting breakdown products
Capillary isoelectric focusing (cIEF)	Primary	Determine charge heterogeneity
Reducing isoelectric focusing	Primary	Determine charge heterogeneity
Western blotting	Primary/secondary/tertiary	Identify immunoreactive fragments of heavy and light chain
Size exclusion chromatography	Tertiary	Determine the presence of aggregates or breakdown products
Differential scanning calorimetry	Tertiary	Verify consistent thermal stability profile
Laser light scattering	Tertiary	Identify aggregates and breakdown products
Microneutralization	Tertiary	Determine the potency of the product
F protein binding ELISA	Tertiary	Determine the potency of the product
Process Impurity Profile		Determined

3.3.2. Host-Dependent Glycosylation

Unique to biological products from living organisms is the presence of carbohydrate chains on the protein. The carbohydrate chains can be attached to the protein either at an asparagine residue (N-linked) or a serine or threonine residue (O-linked). These carbohydrate chains can affect the protein's solubility, how it folds, its in vivo biological function and its pharmacokinetic profile. Identification of these oligosaccharides is an important aspect of biopharmaceutical characterization.

Although the integrity of the protein chain is largely unchanged by the choice of recombinant host, such is not the case with the carbohydrate chains. When a thorough analysis of the glycosylation patterns on these molecules is performed, it is not surprising that more than 30 different oligosaccharides can be present, of varying amounts.

For the mammalian cell lines typically used (Chinese hamster ovary cells or murine cells), the glycosylation observed is similar to that observed with human cells.

Insect cell lines, unlike human cells, cannot sialylate the carbohydrate chains.

The difference in glycosylation by bioengineered plants may be more significant:

> "Glycoproteins produced in plants have two carbohydrate determinants found on the N-glycans of plants and invertebrates, but not found in humans. The first carbohydrate determinate is xylose, a sugar not produced in mammals but commonly found in plants as ß(1,2)-xylose which is linked to ß-mannose; and the second determinant is α-(1,3) fucose, which is attached to glucosamine."[67]

But even within a given expression system, there will also be changes in the carbohydrate chains due to the cell culturing conditions. Lot-to-lot consistency will be important for those biopharmaceutical products that are impacted by the type of carbohydrate present. For example, the degree of galactosylation on the monoclonal antibody rituximab (Rituxan®) affects its biological activity, and lot release testing with a tight specification is required to ensure that the correct ratio of oligosaccharide molecules are present on the molecule for each manufactured lot.[129]

3.3.3. Host-Dependent Impurities

Regardless of the recombinant host chosen to express the biopharmaceutical, standard host-related impurities such as residual DNA or host proteins are considered during the characterization of the biopharmaceutical.

However, the choice of certain recombinant hosts can lead to additional scrutiny during characterization of the process-related impurities. For example, the following process-related impurities need to be part of the characterization of biopharmaceuticals expressed in bioengineered plants[67, 97]

- Secondary plant metabolites (alkaloids, glycosides)

- Toxicants (protease inhibitors, aflatoxins, other mycotoxins)

- Pesticides/herbicides/fungicides

- Toxic metals (in plants that accumulate these)

- Allergens (from contamination with mold, animal dander, animal excrement, or dust mite due to field or storage conditions)

3.3.4. Impact of Molecular Variants on Biological Activity

Analysis of the observed molecular variants is expected. This analysis, at a minimum, involves their identification and an assessment of their bioactivity. At the End of Phase 2 (EOP2) CMC meeting with the FDA, this area becomes one of the topics of interest for discussion.[17]

It is difficult to predict the effect, if any, of molecular variants on the biological activity. Glycosylation for many monoclonal antibodies does not seem to affect biological activity. However, for other monoclonal antibodies, such as Rituxan[®], the biological activity has been demonstrated to be a direct function of the composition of the carbohydrate moieties contained within the molecule.[26] On the other hand, the C-terminal lysine heterogeneity observed on monoclonal antibody heavy chains does not usually impact potency, being so distant from the active portion of the monoclonal antibody. In a careful study of six different human IgG_1 monoclonal antibodies manufactured and characterized by the same company, surprises were noted All six monoclonal antibodies had the same five common sources of heterogeneity:

- Asparagine$_{297}$ oligosaccharides
- C-terminal lysine processing
- Methionine oxidation in the Fc region
- Hinge-region fragmentation
- Glycation of lysine residues

All six monoclonal antibodies were from the same recombinant host (CHO) and all had a similar purification process. Despite these commonalities, each monoclonal antibody gave a different surprise during characterization ranging from issues of deamidation, or sialic acid content or galactosylation.[130]

Molecular variants that have comparable properties to those of the desired product with respect to activity, efficacy and safety, are considered 'product-related substances'. Those molecular variants that do not have comparable properties to those of the desired product with respect to activity, efficacy and safety, are considered 'product-related impurities'.[126]

Furthermore, if it can be demonstrated that the molecular variants are consistent between manufactured lots, and that the lots used in the clinical development program contain the same consistency of impurities, the individual assessment of biological activity may not be necessary:

> "The manufacturer should define the pattern of heterogeneity of the desired product and demonstrate consistency with that of the lots used in preclinical and clinical studies. If a consistent pattern of product heterogeneity is demonstrated, an evaluation of the activity, efficacy and safety (including immunogenicity) of individual forms may not be necessary."[126]

3.4. Characterization of a Gene Therapy Biopharmaceutical

While not as extensive as is required for proteins, the DNA used in plasmid vectors and viral vectors still needs to be thoroughly characterized. Table 65 presents some of the regulatory expectations for this characterization.

Table 65. Characterization of a gene therapy biopharmaceutical[35, 94]

Molecular Properties of the DNA Product	Rigorous characterization of the final processed product or the finished product and its individual components, where appropriate, is essential and their stability should be established by an appropriate range of molecular and biological methods.
	Suitable tests should be included to establish, for example, that complexed nucleic acid has the desired biochemical and biological characteristics required for its intended use. Immunological and immuno-chemical tests may provide valuable information.
	In the case of viral vectors, tests should be included to show integrity and homogeneity of the recombinant genome.
Vector Sequence	Early in product development, vector characterization consisting of sequence data of appropriate portions of vectors and /or restriction mapping supplemented by protein characterization is acceptable
	For later phases of product development and licensure, more extensive sequencing information should be provided. When sequencing of the entire vector is not feasible due to the size of the construct, it may be sufficient to sequence the genetic insert plus flanking regions and any significant modifications to the vector backbone or sites known to be vulnerable to alteration during the molecular manipulations.
Impurity Profile	The purity of the final processed product should be determined and the level(s) of impurities considered as acceptable should be justified. These include, e.g., nucleic acids derived from bacteria used for the production of plasmid DNA, extraneous nucleic acids in vector preparations or other impurities.
Absence of Replication-Competent Virus	In the case of replication-deficient viral vectors, it is essential that all measures/steps are being taken to minimise the possibility that they become contaminated with replication-competent viruses during the production processes. Tests are essential to show that replication-competent viruses are below an acceptable level. In the case of replication-competent adenovirus, the level set should be demonstrated to be safe in appropriate animal and/or human studies.
Consistency	A minimum of 3 successive batches of the bulk product should be characterized as fully as possible to determine consistency with regard to identity, purity, potency and safety. Thereafter, a more limited series of tests may be appropriate.

3.5. Applying the Minimum CMC Continuum to Characterization

The characterization of biopharmaceuticals requires considerable long-term planning. The regulatory agencies permit the start of clinical trials with initial characterization in place, but they also require full characterization by the time of the BLA/NDA filing, as indicated in Figure 15. Those companies that invest adequate resources for product characterization are building a solid foundation for understanding the strengths and weaknesses of the their specific biopharmaceutical.

Figure 15. Minimum CMC continuum for biopharmaceutical characterization

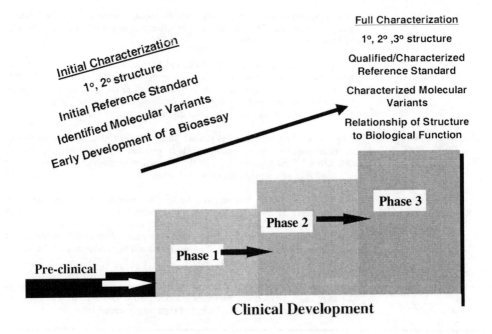

4. RELEASE TESTING AND SPECIFICATIONS

'Conformance to specification' means that the API or drug product, when tested according to the listed release assay by QC, will meet the acceptance criteria. Specifications are critical quality standards that are proposed and justified by the manufacturer, and approved by regulatory agencies as conditions of market approval.

4.1. Regulatory Expectations During Clinical Development

The minimum CMC continuum applies to release testing and specifications for the biopharmaceutical product. At each stage of clinical development, the regulatory agencies expect that the minimum CMC requirements will be met, and that these requirements will increase and tighten as clinical development progresses. This is indicated in the FDA's guidance documents on what they expect to see included in the IND submissions.

4.1.1. Phase 1 IND Submission

API: "A brief description of the test methods used should be submitted. Proposed acceptable limits supported by simple analytical data (e.g.,...HPLC chromatograms to support the purity level and impurities profile) of the clinical trial materials should be provided. Submission of a copy of the certificate of analysis is also suggested. The specific methods will depend on the source and type of drug substance (e.g., ... biotechnology-derived products). Validation data and established specifications ordinarily need not be submitted at the initial stage of drug development. However, for some well-characterized, therapeutic biotechnology-derived products, preliminary specifications and additional validation data may be needed in certain circumstances to ensure safety in Phase 1."[16]

Drug Product: "A brief description of the proposed acceptable limits and the test methods used should be submitted. Tests that should be submitted will vary according to the dosage form. For example, for sterile products, sterility and non-pyrogenicity tests should be submitted. Submission of a copy of the certificate of analysis of the clinical batch is also suggested. Validation data and established specifications need not be submitted at the initial stage of drug development. For well-characterized, therapeutic, biotechnology-derived products, adequate assessment of bioactivity and preliminary specifications should be available." [16]

The FDA recognizes that the amount of information about the release tests and specifications will be very limited in the early stage of clinical development. Preliminary specifications will occur at Phase 1.

4.1.2. Phase 2 IND Submission

API: "Any changes in the specification should be reported. The analytical procedures used to perform a test and to support the acceptance criteria should be indicated (e.g., HPLC). A complete description of the analytical procedure and supporting validation data should be available on request. Any changes in the tentative acceptance criteria should be stated. Test results, analytical data, and certificates of analysis (COA) of clinical trial material prepared since the filing of the original IND should be provided. If degradation of the drug substance (or drug product) occurs during manufacture and storage, this should be considered when establishing acceptance criteria and monitoring quality. "[36]

Drug Product: "Changes to the specification should be reported. Chemical tests (e.g., dissolution, identity, assay, content uniformity, impurities) and microbiological tests (sterility and endotoxin/LAL for sterile products; antimicrobial preservative and microbial limits for non-sterile dosage forms) that were added, deleted, or changed since phase 1 should be indicated. The analytical procedure used to perform a test and support the acceptance criteria should be indicated (e.g., HPLC). The complete description of the analytical procedure and supporting validation data should be available upon request. Any changes in tentative acceptance criteria should be stated for each test performed. Data updates on the degradation profile should be provided so safety assessments can be made. A summary table of the test results, analytical data (e.g., chromatogram), and COA for lots of the drug product used in clinical studies should be provided." [36]

4.1.3. Phase 3 IND Submission

API: "A detailed listing of all the tests performed (e.g., description, identity, assay, loss on drying) should be provided. A general description of the analytical procedures should be provided that includes a citation to the specific USP monograph or general chapter or the sponsor's standard test procedure number, as appropriate. A complete description of the non-USP analytical procedures with appropriate validation information should be provided. The assay validation program should be designed to delineate various analytical parameters such as accuracy, precision, and specificity, as well as detection limits, quantitative limits, linearity, and range, where appropriate. Tentative acceptance criteria should be established for each test performed. Impurities should be identified, qualified, and quantified, as appropriate. Suitable limits based on manufacturing experience should be established. A summary table of updated test results, analytical data (e.g., ... HPLC chromatograms, microbial limits for incoming raw materials prior to sterilization) and COAs for the lots of drug substance used in clinical trials should be provided."[36]

Drug Product: "Updates on the information specified for phases 1 and 2 should be provided. A general description of the analytical procedures used should be provided that includes a citation to the specific USP monograph, general chapter, or the sponsor's standard test procedure number, if appropriate. A full description of the non-USP analytical procedures with appropriate validation information should be provided. The acceptance criteria should be stated for each test performed. Degradation products should be identified and qualified."[36]

By Phase 3, it is expected that the test methods are validated and more appropriate specifications have been established.

4.2. Regulatory Expectations for the BLA/NDA Submission

By the time of filing of the dossier for market approval, the release tests and the assigned specifications are expected to be complete and justified. Table 66 presents the release testing and specification information that is expected to be included in the filing.

Table 66. Release testing and specification information to be included in the BLA/NDA dossier for recombinant proteins and monoclonal antibodies[40, 41, 42]

	API	Drug Product
Specifications and Test Methods	Specifications and tests for the drug substance sufficient to assure its identity, purity, strength and/or potency, as well as lot-to-lot consistency should be submitted.	The sampling procedures for monitoring a batch of finished drug product should be included.
	The analytical procedures should be provided.	A description of all test methods selected to assure the identity, purity, strength and/or potency, as well as the lot-to-lot consistency of the finished product and the specifications used for the drug product should be submitted.
	Certificates of analysis and analytical results for at least three consecutive qualification lots of the drug substance should be submitted.	The analytical procedures should be provided.
	Justification of the specifications should be provided.	Certificates of analysis and analytical results for at least three consecutive batches should be provided.
		Justification of the specifications should be provided.

| Test Method Validation | Validation of the analytical systems and the data should be provided for non-compendial methods to demonstrate the system suitability. | The validation data for system stability for all non-compendial tests should be provided.

If compendial methods have to be validated to ensure non-interference of special inactive ingredients, the results of those validation studies should be submitted. |

This information must be complete in the filing. The test methods must be validated and the specifications justified.

4.3. Appropriate Release Test Methods

The selection of the test methods to be included in lot release of the biopharmaceutical is most important for the manufacturer. Insufficient testing or incorrect choice of test methods, could lead to patient safety concerns. Too much testing could result in extended resource commitments and delays in product release. The selection of the specific test methods is driven by both the type of biopharmaceutical to be tested (e.g., a recombinant protein, a monoclonal antibody, a DNA vector, etc.) and by the test methodology already resident in the QC laboratory.

The test methods selected for release testing must meet regulatory expectations:

"Laboratory controls shall include the establishment of scientifically sound and appropriate ... test procedures ...."[131]

4.3.1. Not All Release Testing is at API or Drug Product Stage

Release testing at in-process stages during the production and purification of the biopharmaceutical, rather than the API or drug product stage, may be appropriate and justifiable. For example, to assure the safety of the product, the most sensitive process stage to test sterility to confirm the absence of microbes, is at the API and drug product stages. However, the most sensitive process stage to test sterility to confirm the absence of adventitious agents such as mycoplasma and virus, is prior to the harvest of the cell culture stage.

When release testing is performed at an in-process stage, it should appear, with a specification, on the Certificate of Analysis (CofA) of the API or the drug product. For example, the drug product CofA for infliximab (Remicade®) lists the following release tests performed at the end of fermentation stage[71]:

Mycoplasma
>   Cultivable
>   Non-cultivable

Retrovirus testing
>   XC plaque assay
>   S$^+$L$^-$ focus assay
>   Dunni cell assay

In vitro test for adventitious viruses on:
>   Vero cells
>   MRC-5
>   HeLa cells
>   Host C168J cells

It should also be noted, that not all QC testing is release testing. Most in-process testing is performed by QC to assess the consistency of the manufacturing process. Such in-process test methods should be under action limits. These action limits, when exceeded, are used to initiate an investigation or to require further action to be taken. However, in-process testing performed at critical decision making steps (e.g., the mycoplasma and virus testing prior to harvest) should be under specifications As any other specification, if it is not met, the lot cannot be released.

4.3.2. Elimination of Release Testing by Process Validation

With the major commitment in process validation by the biopharmaceutical manufacturer, it would be nice to have some daily return on the investment. Such would be the case if process validation could lead to a reduction in the number of tests that need to be performed for lot release.

Process validation rather than release testing may be justifiable. Process validation clearance studies are performed to document the capacity of the purification steps in reducing the impurity content in the product. Based upon the results of those clearance studies, it may be possible to eliminate the need for a lot release assay:

> "For certain impurities, testing of either the drug substance or the drug product may not be necessary and may not need to be included in the specification if efficient control or removal to acceptable levels is demonstrated by suitable studies."[126]

Process validation clearance studies can be performed on most process-related impurities. Testing for residual host cell DNA in the product illustrates this approach:

> "Validation studies (e.g., spiking experiments using an adequate size distribution of DNA) should be performed in an attempt to identify the major steps capable of reducing the DNA burden and to document the capacity of those steps in reducing residual cellular DNA content n the final product, to an acceptable and defined level. The technology for DNA quantitation is now well defined and reproducible.

> In addition to the validation studies, results of DNA quantitation on a minimum number of production batches (e.g., 5 consecutive batches) should be provided to demonstrate the reproducibility of the production process in reducing residual DNA to the level expected from the validation studies.

> Based on satisfactory validation data and consistent results on a limited number of production batches, it seems reasonable not to perform routinely host cell DNA tests at the purified bulk level (or other appropriate steps)."[108]

But the process validation clearance approach cannot always be used. For example, if the residual DNA to be measured is in a biopharmaceutical expressed using a recombinant organism that the regulatory agency may have limited experience with or if the residual DNA is from transforming DNA sequences from viral vectors, then the regulatory agencies may still require lot release testing.[108]

Also, the process validation clearance approach may receive different reception at different regulatory agencies. For example, the use of this process validation approach for eliminating the need to perform residual host cell protein tests on lot release is generally acceptable to the FDA, but EMEA takes a more conservative approach.[108]

4.3.3. Test Method Parameters Required for Release: Proteins

Release testing is performed first on the API, then on the drug product. Table 67 presents the general regulatory guidance on which product parameters need to be tested for lot release of recombinant proteins and monoclonal antibodies:

Table 67. Test method parameters required for release of both APIs and drug products of recombinant proteins and monoclonal antibodies[29, 126, 132]

Appearance and Description	A qualitative statement describing the physical state (e.g., solid, liquid) and color.
	For a drug product, clarity is also described.
Identity (21 CFR 610.14)	Should be highly specific and should be based on unique aspects of its molecular structure and/or specific properties.
	More than one test may be necessary to establish identity.
	The test can be qualitative in nature.
Purity (21 CFR 610.13)	The absolute purity of biotechnological and biological products is difficult to determine and the results are method-dependent. Consequently, the purity is usually estimated by a combination of methods.
	Both process-related and product-related impurities need to be considered.
	For the drug product, if impurities are qualitatively and quantitatively (i.e., relative amounts and/or concentrations) the same as in the API, testing of these impurities is not necessary.
	The choice and optimization of analytical procedures should focus on the separation of the desired product and product-related substances from impurities.
	Endotoxin measurement is required the drug product. The LAL assay may be used when validated against the rabbit pyrogen assay.
Potency (21 CFR 610.10)	A relevant, validated potency assay should be part of the specifications for a biotechnological and biological product.
	When an appropriate potency assay is used for the drug product, an alternative method (physicochemical and/or biological) may suffice for quantitative assessment of the API; when an appropriate potency assay is used for the API, an alternative method (physicochemical and/or biological) may suffice for quantitative assessment of the drug product. However, the rationale for such a choice must be provided.
Sterility (21 CFR 610.12)	Bioburden is acceptable for the API.
	Sterility is required for injectable drug products; microbial content assay is allowed for topical drug products
Quantity	Quantity is usually based on protein content (mass), using an appropriate assay.
	In cases where product manufacture is based on potency, there may be no need for an alternate determination of quantity.
General Tests	Physical description and the measurement of other quality attributes (e.g., pH and osmolarity) are often important.
Dosage Forms	Certain unique dosage forms may need additional tests.

The regulatory guidance for biopharmaceuticals also refers to the need to perform the General Safety Test (21 CFR 610.11), which requires injecting the drug product into both mice and guinea pigs.[133]   For specified biologics (which include the recombinant proteins and the monoclonal antibodies), this code requirement has been waived.[134]

Several examples from regulatory agency websites illustrate appropriate lot release test method strategies for recombinant proteins and monoclonal antibodies. Table 68 illustrates different lot release test method strategies for the API.

Table 68. Comparison of reported lot release testing strategies for two biopharmaceutical APIs[135, 136]

Release Test Parameter	Interferon beta-1a (Rebif®)	Monoclonal Antibody Daclizumab (Zenapax®)
Appearance and Description	Appearance	Appearance
Identity (21 CFR 610.14)	Peptide map	Peptide map
Purity (21 CFR 610.13)	SEC-HPLC IEF SDS-PAGE	SEC-HPLC IEF SDS-PAGE Western blotting
	N-terminal amino acid sequence RP-HPLC Glycan Content Endotoxin (LAL) Residual host cell DNA Residual host cell proteins	N-Terminal amino acid sequence
		Endotoxin (LAL) ($\leq$ 2.5 EU/mg) Residual host cell DNA Residual host cell proteins
Potency (21 CFR 610.10)	Cytopathic effect bioassay	Cell growth inhibition bioassay Receptor binding assay
Sterility (21 CFR 610.12)	Bioburden	Bioburden
Quantity	RP-HPLC	UV
General Tests	pH	pH Osmolality

Table 69 (for a liquid drug product) and Table 70 (for a lyophilized drug product) provide insight into the strategy of designing a lot release test program that applies to both the API and the drug product. Many release tests are performed on both the API and the drug product (e.g., potency which is typically expressed as units/mg in API and units/mL in drug product). Some release tests are performed only on either the API or the drug product (e.g., process-related impurities typically performed on API but particulate formation performed on drug product). Some release tests are modified between the API and drug product (e.g., bioburden for the API but sterility for the drug product), because of more stringent safety requirements expected at the drug product stage. Some release tests are completely different between the API and drug product (e.g., identity by peptide map for the API but identity by cIEF for the drug product), frequently due to convenience of use.

Table 69. Reported lot release testing strategy for the monoclonal antibody, rituximab (Rituxan®)[120]

Release Test Parameter	API Release Tests	Drug Product Release Tests
Appearance and Description	Color and Appearance	Color and Appearance Particulate Matter
Identity (21 CFR 610.14)	Peptide Map	Capillary IEF
Purity (21 CFR 610.13)	SEC-HPLC Fragment IEC-HPLC SDS-PAGE Glycan Content Endotoxin (LAL)	SEC-HPLC Endotoxin (LAL)
Potency (21 CFR 610.10)	Complement dependent cytotoxicity bioassay	Complement dependent cytotoxicity bioassay
Sterility (21 CFR 610.12)	Bioburden	Sterility
Quantity	UV	UV Container volume
General Tests	pH Polysorbate concentration	pH Osmolarity

**Table 70. Reported lot release testing strategy for the monoclonal antibody,
infliximab (Remicade®)[71]**

Release Test Parameter	API Release Tests	Drug Product Release Tests
Appearance and Description	Appearance	Appearance Color after reconstitution Visible particles after reconstitution Turbidity after reconstitution Reconstitution time
Identity (21 CFR 610.14)	Gel filtration-HPLC	Gel filtration-HPLC IEF Immunodifffusion assay
Purity (21 CFR 610.13)	Gel filtration- HPLC SDS-PAGE Endotoxin (LAL)	Gel filtration- HPLC SDS-PAGE Endotoxin (LAL)
Potency (21 CFR 610.10)	Cell growth inhibition bioassay	Cell growth inhibition bioassay
Sterility (21 CFR 610.12)	Bioburden	Sterility
Quantity	UV	UV Uniformity of dosage units
General Tests	pH	pH Residual moisture

4.3.4. Test Method Parameters Required for Release: DNA Vectors

Release testing is performed first on the API, then on the drug product. Table 71 presents the general regulatory guidance on which product parameters need to be tested for lot release of gene therapy biopharmaceuticals .

Table 71. Test method parameters required for release of both APIs and drug products of gene therapy vector biopharmaceuticals[35, 132]

Identity (21 CFR 610.14)	Test for vector identity by methods such as restriction enzyme mapping with multiple enzymes or PCR should be performed on the API. In case of a facility making multiple constructs, it should be verified that the identity testing is capable of distinguishing the constructs and detecting cross-contamination.
Purity (21 CFR 610.13)	Test for total DNA or RNA content if appropriate to vector composition (e.g., A_{260}/A_{280}). Test for homogeneity of size and structure, supercoiled vs. linear (e.g., agarose gel electorphoresis) Test for contamination with RNA or with host DNA (e.g., gel electrophoresis, including test with bacterial host-specific probe). Test for proteins if present as a contaminant (e.g., silver stained gel). Test for non-infectious virus in cases in which that would be a contaminant, such as empty capsids. Tests for toxic materials involved in production. Endotoxin measurement is required the drug product. The LAL assay may be used when validated against the rabbit pyrogen assay.
Potency (21 CFR 610.10)	Whenever possible, a potency assay should measure the biological activity of the expressed gene product, not merely its presence. If no quantitative potency assay is available, then a qualitative potency test should be performed.
Sterility (21 CFR 610.12)	Sterility is required.
Adventitious Agents	Mycoplasma testing is required, as specified in the Points to Consider in the Characterization of Cell Lines Used to Produce Biologicals (1993). Testing for an appropriate range of possible contaminating viruses is recommended.
Replication Competent Virus	Specific testing requirements for demonstrating the absence of replication competent viruses (retroviruses and adenoviruses).

The regulatory guidance for biopharmaceuticals also refers to the need to perform the General Safety Test (21 CFR 610.11), which requires injecting the drug product into both mice and guinea pigs.[133] For specified biologics (which include the DNA plasmid gene therapy products), this code requirement has been waived.[134] But, for the virus vector gene therapy products, this test is still a requirement.

4.4. The Bioassay – Absolute Requirement for a Biopharmaceutical

A bioassay is an analytical procedure that uses a biological response test method. Bioassays are used to determine the potency, the biological activity, of a biopharmaceutical:

> "The word potency is interpreted to mean the specific ability or capacity of the product … to effect a given result."[137]

A correlation between the expected clinical response and the bioassay is expected to be established:

> "It is desirable that the assay(s) bear the closest possible relationship to the putative physiologic/pharmacologic activity of the product and be sufficiently sensitive to detect differences of potential clinical importance in the function of the product."[29]

Typical bioassays include:

- Animal-based biological assays, which measure an organism's biological response to the product

- Cell-based biological assays, which measure biochemical or physiological response at the cellular level

- Biochemical assays, which measure biological activities such as enzymatic reaction rates or biological responses induced by immunological interactions

Development of a bioassay takes considerable time and resources. For many biopharmaceuticals, Phase 1 is initiated either without a bioassay or with the use of only an immunoassay. However, at Phase 2, the regulatory agencies expect a true bioassay to be under development. The appropriateness of the chosen bioassay and the some details of its validation (e.g., specificity and precision) become a major discussion point at the End-of-Phase 2 CMC meeting.[17] At Phase 3, the expectation is that the bioassay will be undergoing full validation. For the BLA/NDA submission, the bioassay is expected to be validated.

Bioassays don't have to be complex, highly variable test methods. A number of manufacturers are developing cell-based bioassays that are simple, have high throughput (performed in 96-well microtiter plates) and are reproducible.[138]

It is possible to use an immunoassay, the binding of the product to a receptor or antibody, to measure the biological activity of a biopharmaceutical. But, binding to a receptor or an antibody, is not necessarily causing a biological effect. To meet the requirement of a biological assay for market approval, a rigorous correlation between the immunoassay and biological function would have to be demonstrated:

> "Immunoassays of interferon potency offer the potential advantages of improved ease of performance, sensitivity, and consistency. However, the suitability of such assays for routine potency testing must be established individually in any laboratory desiring to do so, by demonstration that potency results obtained through immunoassays consistently correlate with biological activity under a variety of circumstances."[1]

It is with the therapeutic monoclonal antibodies that immunoassays have found acceptance as bioassays. For therapeutic monoclonal antibodies, if one can demonstrate that the mechanism of action for the biopharmaceutical is based solely on binding, then a binding assay is an appropriate bioassay. Such was demonstrated for the monoclonal antibody palivizumab (Synagis®). For this monoclonal antibody, complement played no significant role in neutralizing RSV and Fab fragments were capable of efficiently binding to virus. Antibody preparations with varying degrees of glycosylation were also evaluated and found to play no significant role in virus neutralization. During the course of potency assay development, a protein binding ELISA was demonstrated to be comparable to various biological assays. Collectively these results enabled the use of the binding ELISA for lot release.[139]

It is even possible to use a physicochemical test to measure the biological activity of a biopharmaceutical, but only if the following two parameters are both met[126]:

1. Sufficient physicochemical information about the drug, including higher-order structure, can be thoroughly established by such physicochemical methods, and relevant correlation to biologic activity demonstrated

2. There exists a well-established manufacturing history

The biological activity for recombinant human insulin is one example of this, in which the bioactivity of the molecule is determined by HPLC.[140] However, for most biopharmaceuticals, while the possibility exists, these requirements are most difficult to achieve.

4.5. Test Method Validation – How Much and When?

There are four main documents describing the expectations and procedures for test method validation:

- ICH Q2A: Validation of Analytical Procedures[141]

- ICH Q2B: Validation of Analytical Procedures – Methodology[142]

- FDA Analytical Procedures and Methods Validation[143]

- USP <1225> Validation of Compendial Methods[144]

4.5.1. Regulatory Expectations for Test Method Validation

Regulatory agencies recognize that the analytical methodology will be evolving during the course of clinical development. They also recognize that while the test method is under development, and while the test method is not defined and locked in, it is not possible for the test method to be validated

It for this reason that the regulatory agencies provide some flexibility in the requirement for test method validation at the earlier clinical development stages:

"While analytical methods performed to evaluate a batch of API for clinical trials may not yet be validated, they should be scientifically sound."[22]

"As processes may not be standardized or fully validated, testing takes on more importance in ensuring that each batch meets its specifications"[49]

However, by the time of submitting the BLA/NDA, test method development is expected to be completed and the test method validated:

"Analytical methods should be validated ...."[22]

"All analytical procedures should be fully developed and validation completed when the NDA, ANDA, BLA, or PLA is submitted." [144]

There are 9 validation characteristics that must be considered for each test method:

- Specificity: the ability to assess unequivocally the analyte in the presence of components which may be expected to be present

- Accuracy: the closeness of agreement between the value which is accepted either as a conventional true value or an accepted reference value and the value found

- Precision: the closeness of agreement (degree or scatter) between a series of measurements obtained from multiple sampling of the same homogeneous sample under the prescribed conditions

  Repeatability: intra-assay precision

  Intermediate Precision: inter-assay precision

  Reproducibility: lab-to-lab precision

- Detection Limit: the lowest amount of analyte in a sample which can be detected but not necessarily quantitated as an exact value

- Quantitation Limit: the lowest amount of analyte in a sample which can be quantitatively determined with suitable precision and accuracy

- Linearity: the ability (within a given range) to obtain test results which are directly proportional to the concentration (amount) of analyte in the sample

- Range: the interval between the upper and lower concentration (amounts) of analyte in the sample (including these concentrations) for which it has been demonstrated that the analytical procedure has a suitable level of precision, accuracy and linearity

- Robustness: a measure of the capacity to remain unaffected by small, but deliberate variations in method parameters and provides an indication of reliability during normal usage

Biopharmaceutical test methods are more challenging to validate than the typical test methods used for chemically-synthesized drugs. However, more manufacturers are publishing on their test method validation to provide insight on how to accomplish it: validation of peptide mapping for identity and genetic stability[145, 146], cell proliferation bioassay.[147]

Many biopharmaceutical test methods, unlike the test methods for chemically-synthesized drugs, are not compendial. It should be noted that the United States Pharmacopeia (USP) has been adding more and more biopharmaceutical-related test methods into its official compendia of standards; for example, amino acid analysis, capillary electrophoresis, isoelectric focussing, peptide mapping, polyacrylamide gel electrophoresis and total protein assay. However, it is important to note, that these test methods are in the 'General Information' section of the compendia and have no regulatory binding.[148]

4.5.2. Assay Qualification During Clinical Development

So how much test validation is needed during the clinical development stages? Many QC labs refer to 'test method qualification' when discussing assay needs during Phase 1 and 2.

Before a test method can be validated, a number of important steps must be accomplished first:

- Development of a new test method (if one already is not available) or adaptation of an existing test method to the biopharmaceutical product

  While there may seem be 'off the shelf' analytical test methods for biopharmaceuticals, such as SDS-PAGE or IEF, each test method has to be adapted to the optimum conditions for the specific product (e.g., the gel gradient level in SDS-PAGE and the pH range for the IEF gel). For many biopharmaceuticals, the bioassay may need to be developed from scratch.

- Qualification (IQ/OQ) of the laboratory instrumentation

  Documented steps have to be taken to ensure that the instruments have been properly installed and working as intended by the vendor. Calibration of the proper controls for use of the laboratory instrumentation is important.[149, 150]

- Qualification of critical assay reagents

  Many biopharmaceutical assays depend on reagents and materials that are complex, labile, or derived from processes that can yield variations in product attributes. When these assays are used, the impact of the critical reagents on routine, reliable assay performance should be evaluated. Procedures need to be developed in which each new lot of critical reagent is examined for suitability in the assay.[151]

  Useable shelf life for critical reagents needs to be determined.[143]

- Development of a reference material based on the current manufacturing process

  Reference material from the current manufacturing process needs to be prepared and will be used in side-by-side comparison with product lots. As the manufacturing process is developed, new reference materials made need to be prepared. The final reference material will be thoroughly characterized and included in the BLA/NDA.

The reference material is typically used as a system suitability test for many biopharmaceutical assays (e.g., the comparison of electrophoretic or chromatographic patterns) to demonstrate that the assay system is working properly at the time of analysis.

- Development of an appropriate potency reference standard

The development of the potency reference standard is most challenging. For biological assays, lab-to-lab variation is quite large, which has been documented in several collaborative international studies. For example, in the international collaborative study to assign the infectious titer to an adenovirus Type 5 reference material, 17 laboratories were involved, and their assigned measurements ranged from about 4×10^{10} IU/mL to about 4×10^{10} IU/mL, and entire log range.[152]

For some biopharmaceuticals, primary potency reference standards are now available (the National Institute for Biological Standards and Controls, NIBSC, provides several recombinant protein potency reference standards[153]; the American Type Culture Collection, ATCC, provides an adenovirus potency reference standard for gene therapy virus testing[154]). But for most biopharmaceuticals, there is no established reference standard, so a manufacturer will need to develop its own internal potency standard.

Assay qualification is performing some, or all, of the important steps preceding assay validation. Assay qualification can also include some limited assay validation (e.g., specificity and repeatability).

4.5.3. Applying the Minimum CMC Continuum to Test Method Validation

Having adequate test method qualification early in the clinical development process can yield valuable information about the performance of the test method and its proper control. Waiting to complete test method validation until later clinical development stages, conserves the limited resources in a biopharmaceutical company. However, the risk that the company incurs by waiting to complete the test method validation until later clinical development stages is that it may be discovered too late that the test method is inappropriate or inadequate for its intended use.

There is one category of release test methods that should be validated even at Phase 1: the test methods confirming the safety of the drug product (e.g., LAL for endotoxin and sterility). Patients on clinical trials deserve as much protection as those who purchase the drug from a subscription after it is one the market.

Figure 16 illustrates the minimum CMC continuum for test method validation.

Figure 16. Minimum CMC continuum for test method validation

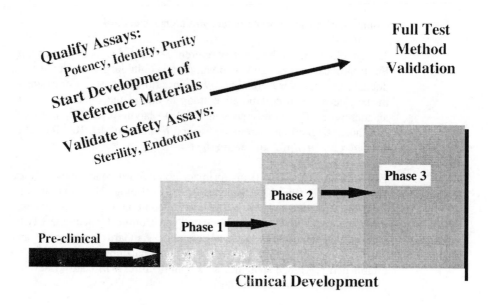

4.6. The Art of Setting a Specification

Assigned specifications need to be scientifically sound and appropriate:

"All specifications ... should be scientifically sound and appropriate ...."[22]

"Laboratory controls shall include the establishment of scientifically sound and appropriate specifications ...."[155]

4.6.1. Development of a Specification

As shown in Figure 17, specification development has an orderly progression. The process begins at Phase 1 where the requirement is for 'preliminary specifications'.[16]

During clinical development, there is the regulatory expectation that these preliminary specifications will be periodically re-examined and considered for change.[36] At the time of the BLA/NDA filing, the final specifications are to be set and justified:

> "Final specifications for the drug substance and drug product are not expected until the end of the investigational process."[34]

After market approval, there is the regulatory requirement of an annual assessment to determine if the approved specifications are still appropriate:

> "Written records required by this part shall be maintained so that the data therein can be used for evaluating, at least annually, the quality standards of such drug product to determine the need for changes in drug product specifications ...."[156]

Figure 17. Development of a specification

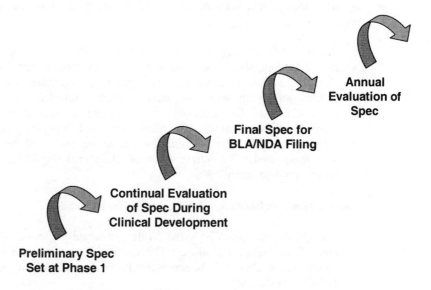

The challenge in setting specifications is not to rush this orderly progression and lock in final specifications before they are required to be finalized.

Even when the final specifications have to be set, there is the danger of assigning the specification based on what a company perceives the FDA reviewers want to see, rather than using a scientific justification for the data available. While it is commendable to set a specification so that it would be accepted by the regulatory agency without causing any delays in market approval, the company pays the price after product approval when the manufactured product is not able to consistently meet an overly tight specification.

According to the ICH Q6B guideline on 'Test Procedures and Acceptance Criteria for Biotechnological/Biological Products', the following points should be taken into consideration when establishing scientifically justifiable final specifications[126]:

- Specifications are linked to a manufacturing process

  "Specifications should be based on data obtained from lots used to demonstrate manufacturing consistency. Linking specifications to a manufacturing process is important, especially for product-related substances, product-related impurities and process-related impurities. Process changes and degradation products produced during storage may result in heterogeneity patterns which differ from those observed in the material used during preclinical and clinical development. The significance of these alterations should be evaluated."

- Specifications should account for the stability of drug substance and drug product

  "Degradation of drug substance and drug product, which may occur during storage, should be considered when establishing specifications. Due to the inherent complexity of these products, there is no single stability-indicating assay or parameter that profiles the stability characteristics. Consequently, the manufacturer should propose a stability-indicating profile. The result of this stability-indicating profile will then provide assurance that changes in the quality of the product will be detected. The determination of which tests should be included will be product-specific."

- Specifications are linked to preclinical and clinical studies

  "Specifications should be based on data obtained for lots used in pre-clinical and clinical studies. The quality of the material made at commercial scale should be representative of the lots used in preclinical and clinical studies."

- Specifications are linked to analytical procedures

  "Critical quality attributes may include items such as potency, the nature and quantity of product-related substances, product-related impurities, and process-related impurities. Such attributes can be assessed by multiple analytical procedures, each yielding different results. In the course of product development, it is not unusual for the analytical technology to evolve in parallel with the product. Therefore, it is important to confirm that data generated during development correlate with those generated at the time the marketing application is filed."

Each of these four points, taken together, allows a manufacturer to set a scientifically sound final specification in their filed BLA/NDA.

4.6.2. Release Versus Shelf-Life Specifications

Depending upon the stability profile for the biopharmaceutical, a manufacturer may set both a release limit (an internal specification used by the manufacture to release a product lot; a specification tighter than that used for the shelf-life) and a shelf-life limit. While regulatory agencies allow this practice, the FDA considers only the limit that applies throughout the shelf-life of the product as the specification:

"The concept of different acceptance criteria for release vs. shelf-life specifications applies to drug products only; it pertains to the establishment of more restrictive criteria for the release of a drug product than are applied to the shelf-life. Examples where this may be applicable include assay and impurity (degradation product) levels. In Japan and the United States, this concept may only be applicable to in-house criteria, and not to the regulatory release criteria. Thus, in these regions, the regulatory acceptance criteria are the same from release throughout shelf-life; however, an applicant may choose to have tighter in-house limits at the time of release to provide increased assurance to the applicant that the product will remain within the regulatory acceptance criteria throughout its shelf-life. In the European Union there is a regulatory requirement for distinct specification for release and for shelf-life where different."[126]

4.6.3. Are There Required Purity/Impurity Limits?

The regulatory agencies have not set any required purity specification value for biopharmaceutical products. However, the regulatory agencies are aware that most biopharmaceuticals can be manufactured to a reasonably high purity level:

"In general, > 95% of the protein should be the expected interferon species."[1]

"Although recombinant DNA products may be demonstrated to be 99% pure by physicochemical characterization, ..."[2]

But to take this awareness and consider it a regulatory requirement is dangerous, because many biopharmaceutical products cannot meet such a high standard, especially at the Phase 1 stage.

Frequently for Phase 1, purity will simply be a 'reported value' from whatever assays are considered purity-indicating. At Phase 1, with limited manufactured lots, and probably not much qualification of the test method available, it would be inappropriate to set a minimum tolerance limit. However, with more manufacturing experience and test method development completed by Phase 2 or 3, then a lower limit for the purity specification would be appropriate. For the BLA/NDA filing, a scientifically justifiable lower limit should be set for the purity specification.

For chemically-synthesized drugs, there is an ICH guidance document that provides a specification decision roadmap for impurities in these products. For example, for an API whose drug product dosing is $\leq$ 2 g/day, the following roadmap applies: If the impurity level in the product is above 0.05%, it needs to be reported; if the impurity level is above 0.10%, it must be identified; and if the impurity level is above 0.15%, the impurity must be qualified.[157] However, no such roadmap exists for biopharmaceuticals. Impurities are handled on a case-by-case approach, in which the manufacturer ultimately defends the impurity profile and levels present in the product.

However, there are four impurities that have identified regulatory target values: residual host cell DNA, residual animal serum, bacterial endotoxin and residual moisture.

The regulatory target value for host cell DNA, which has evolved over the years, is as follows:

Residual Host Cell DNA Target	Timeframe
$\leq$ 10 pg DNA per dose	1980s[2]
$\leq$ 100 pg DNA per dose	1990s[29, 80]
$\leq$ 10 ng DNA per dose	2000s[158, 159]

As the regulatory agencies gained more experience with the biopharmaceuticals, and after they obtained additional information that suggested that host cell DNA may be less tumorigenic than originally thought, they became more comfortable with the amount of residual host cell DNA allowed in the product. This is one example in which the regulatory agencies actually loosened a target expectation, rather than tightened one, over time.

The second regulatory target value is for residual animal serum in the product. Since animal serum may produce allergic responses in human subjects, the regulatory requirement is the residual amount of serum or additives in the final product should not exceed 1:1,000,000 (i.e., 1 ppm).[160]

The third regulatory target value is for bacterial endotoxin. Endotoxin when released into the circulatory system is the initiating event of fever. The maximum endotoxin dose for an injectable drug is set at 5 endotoxin units (EU) per kg of patient weight.[161] For a 'typical patient', of say 70 kg, this would correspond to a maximum endotoxin load of 350 EU per dose. However, many of the patients receiving biopharmaceuticals are sick and have much less weight. If the biopharmaceutical is being injected into newborn infants, in which the patient weight is only 2 kg, the maximum endotoxin load becomes only 10 EU per dose. This oversight has occurred with chemically-synthesized drug products, and in one case, was believed responsible for the deaths of 20 newborns in Brazil.[162] It is one of the reasons that biopharmaceutical manufacturers strive to achieve a low value for residual bacterial endotoxin in the drug product.

The fourth regulatory target is for residual moisture in lyophilized products. Each lot of lyophilized product must be tested for residual moisture.[163] The general recommendation for most biological products is that the residual moisture should not exceed 1.0 percent by the gravimetric method.[164]

In addition, for one type of impurity, the use of antibiotics in the manufacturing process, while there is no regulatory target, there is a regulatory requirement to report the type and amount on the product label.[165]

4.6.4. Strategic CMC Tips for Setting Specifications

While most biopharmaceutical manufacturers do not reveal their strategy for setting specifications, it is possible to get a glimpse at how it is done from publications and regulatory reviews.

Most specifications evolve during clinical development. They typically evolve from an empirical approach (lower limit or broad range) based on very little manufacturing experience and data to a statistical approach (typically mean +/- 3 standard deviations) based on, hopefully, much manufacturing experience and data.

Such a strategy is used for the potency specification1[166]:

Clinical Stage	Assigned Specification for API	Example
Phase 1	report measured value in units	_ 10^8 U/mg
Phase 2	measured value must be greater than a set lower limit	$\geq 1 \times 10^8$ U/mg
Phase 3	measured value must be within a set range	1–10 10^8 U/mg
BLA/NDA	scientifically justifiable, calibrated against international standard; results reported in international units (IU)	3.3-8.5 10^6 IU/mg

It needs to be strongly emphasised that for the safety specifications, there is no evolution during the clinical development stage. The limits must meet the regulatory requirements even at Phase 1. Thus, for example, the sterility specification at Phase 1 (USP requirement of 'no growth') is the same as that for market approval (USP requirement of 'no growth').

For some specifications, especially when they directly impact the clinical safety, non-clinical considerations must be incorporated into the final assigned specification. For example, the testing summarized in Table 72 was done to evaluate the appropriate concentration range for sodium dodecyl sulfate (SDS) in recombinant interleukin-2:

Table 72. Test results from rIL-2 preparations with varying SDS concentrations[102]

SDS (ug/mg)	Turbidity (cm³/g)	Pelletable Protein (%)	Bioactivity (x10⁶ IU/mg)	Plasma IL-2 at 1 hr (ng/mL)	Mortality of BDF1 Mice Bearing Lung Metastases
25	0.9	97	20	< 1	Not tested
65	5.1	2	18	12	Increased mortality
95	1.0	1	18	16	Not tested
125	0.6	1	21	24	Increased mortality
157	0.6	2	20	26	Minimal mortality (tested at 165 ug/mg)
197	0.3	1	22	30	Minimal mortality (tested at 204 ug/mg
1086	0.03	1	16	23	Increased mortality

SDS concentrations below approximately 100 ug/mL were associated with increased turbidity and lower plasma levels. SDS concentrations between 165 and 204 ug/mg induced minimal mortality to the mice. Based on these physicochemical and biological, as well as non-clinical studies, the final specification for rIL-2 was set at 160–195 ug SDS/mg IL-2.

How much data is 'enough' when setting a final specification? This is a major problem for biopharmaceutical companies, when they must more fast through the clinical development phases and may not require many production runs to achieve the clinical supplies needed for the studies. From my own experiences, a decade ago, it was typical for a manufacturer to have more than 30 production lots manufactured during clinical development to establish scientifically sound specifications for the market approval dossier. Today, many manufacturers may have only produced 5 lots during clinical development. But is it scientifically justifiable to set final specifications in the BLA/NDA with such limited manufacturing lot experience? The FDA is aware of this concern, and has employed the use of 'interim regulatory specifications':

> "Occasionally, an applicant may wish to propose *interim acceptance criteria* for a specific test because there is some uncertainty whether the same type of results will continue to be observed for production batches. This uncertainty often occurs when (1) there are limited data available at the time the application is submitted and/or (2) the manufacturing process for production batches will be different (e.g., scale, equipment, site) from that used to produce the batches used to support the application and the effect, if any, of the differences has yet to be characterized.
>
> The proposal should include the (1) reason why the interim acceptance criteria are being proposed, (2) number of consecutive production batches that will be produced and tested and/or the time frame before the acceptance criteria will be finalized, (3) data analysis plan, and (4) proposed reporting mechanisms for finalizing the acceptance criteria when the proposed final acceptance criteria are tighter, broader, or the same as the interim acceptance criteria. An applicant should not propose using interim acceptance criteria as a substitute for providing recommended or agreed upon (e.g., at pre-NDA meetings) information in an application.
>
> For NDAs, finalization of interim acceptance criteria will be a Phase 4 commitment."[42]

Recent biopharmaceutical product approvals have used this interim regulatory specification approach, as shown in Table 73.

Table 73. Use of interim regulatory specifications for biopharmaceuticals with a Phase 4 commitment to finalize[167, 168, 169, 170, 171]

Approved Biopharmaceutical	FDA Approval Date	Phase 4 Specification Commitment
Drotrecogin alfa (Xigris™)	November 21, 2001	"The drug substance release specifications will be re-evaluated and submitted to the BLA by May 1, 2002."
		"The drug product release specifications will be re-evaluated and submitted to the BLA by February 1, 2004."

Darbepoetin alfa (Aranesp™)	September 17, 2001	"To set quantitative limits for the 'Mapping of N-glycosylated Sialylated Oligosaccharides' following manufacture of the first 30 commercial lots." "This decision was to allow Amgen to use limits encompassing 4 standard deviations, with a Phase IV commitment to narrow the limits to 3 SDs when sufficient manufacturing history with commercial lots has been accumulated."
	July 19, 2002	"To submit revised release specifications for the amount of oxidized Darbepoetin alfa in the final Albumin (Human) formulated product based upon the analysis of data obtained from 30 commercial lots of final material by July 30, 2003."
Laronidase (Aldurazyme®)	April 30, 2003	"To re-evaluate drug substance and drug product specifications after a minimum of 30 additional lots of Laronidase are produced. A report of the results will be submitted to the FDA by February 28, 2005."

5. BIOPHARMACEUTICAL STABILITY AND EXPIRATION DATING

Molecular conformation of a biopharmaceutical is dependent on noncovalent as well as covalent forces, making these products particularly sensitive to environmental factors such as temperature changes, oxidation, light, and shear. It is through a well designed stability program that a manufacturer is confident in their storage conditions for the biopharmaceutical.

Manufacturers must have an ongoing stability program to support their biopharmaceutical product(s). Stability will need to be determined for each of the following seven areas:

- In-process hold steps (e.g., harvested fluids or cell paste, column eluates)

- API

- Shipment/transport of API (if applicable)

- Intermediates (e.g., chemical modification or conjugation of API)

- Drug product

- In use product (e.g., hold time after reconstitution of a lyophilized product), where appropriate

- Shipment/transport of drug product (from manufacturer to distributor to patient)

5.1. Regulatory Expectations During Clinical Development

The minimum CMC continuum applies to stability testing and expiration dates for the biopharmaceutical product. At each stage of clinical development, the regulatory agencies expect that the minimum CMC requirements will be met, and that these requirements will increase as clinical development progresses:

> "For example, although stability data are required in all phases of the IND to demonstrate that the new drug substance and drug product are within acceptable chemical and physical limits for the planned duration of the proposed clinical investigation, if very short-term tests are proposed, the supporting stability data can be correspondingly limited."[34]

5.1.1. Phase 1 IND Submission

For the Phase 1 IND, the amount of stability information needed is minimal, but stability data must be provided:

> API: "A brief description of the stability study and the test methods used to monitor the stability of the drug substance should be submitted. Preliminary tabular data based on representative material may be submitted. Neither detailed stability data nor the stability protocol should be submitted."[16]
>
> Drug product: "A brief description of the stability study and the test methods used to monitor the stability of the drug product packaged in the proposed container/closure system and storage conditions should be submitted. Preliminary tabular data based on representative material may be submitted. Neither detailed stability data nor the stability protocol should be submitted."[16]

Also, inherent in the stability data provided, is the commitment by the manufacturer to place the first clinical lots onto a defined stability program.

5.1.2. Phase 2 IND Submission

By Phase 2, the expectation is that clinical lots have been placed on stability and that the stability program has been further refined:

API: "Performance of stability stress studies with the drug substance early in drug development is encouraged, as these studies provide information crucial to the selection of stability-indicating analytical procedures for real-time studies.

A stability protocol should be submitted that includes a list of tests, analytical procedures, sampling time points for each of the tests and the expected duration of the stability program. Preliminary stability data based on representative material should be provided. All stability data for the clinical material used in the phase 1 study should be provided."[36]

Drug Product: "A stability protocol should be submitted that includes a list of the tests, analytical procedures, sampling time points for each of the tests, and the expected duration of the stability program. As in phase 1, the stability of the reconstituted solution, when applicable, should be studied and data provided. Preliminary stability data should be based on representative material. All available stability data for the clinical material used in phase 1 should be provided.

Stress testing (e.g., photostability) on the drug product should be conducted."[36]

Stress testing is now required in order to allow the manufacturer select those stability test methods that are stability-indicating.

5.1.3. Phase 3 IND Submission

By Phase 3, the FDA wants to see the details of the defined stability protocol and wants to understand the stability program design:

API: "If not performed during phase 2 studies, stress studies should be conducted to demonstrate the inherent stability of the drug substance, potential degradation pathways and the capability and suitability of the proposed analytical procedures. The stress study should assess the stability of the drug substance at various pH solutions, in the presence of oxygen and light, and at various elevated temperature and humidity increments.

The stability protocol submitted should include a detailed description of the drug substance under investigation, its packaging, a list of the tests to be conducted, analytical procedures to be used, sampling time points for each test, temperature/humidity conditions to be studied, and the expected duration of the accelerated and long-term testing program. Tabulated data should be presented and should include the lot number, manufacturing site, and the date of manufacture of the drug substance lot. Each table should contain data from only one storage condition. Individual data points for each test should be reported. Representative chromatograms and spectra should be provided, when applicable.

A short description should be provided for each parameter being investigated in the stability program studies (i.e., stress, long-term, and accelerated studies) demonstrating that appropriate controls and storage conditions are in place to ensure the quality of the drug substance used in clinical trial. Test unique to the stability program should be adequately defined and described."[36]

<u>Drug Product</u>:   "One-time stress studies should be conducted and results demonstrating the inherent stability of the drug product, potential degradation products, and the capability of the analytical procedures should be included. These one-time studies should also examine the stability of the drug product in the presence of light.

The stability protocol should include a description of the drug product under investigation in the stability program, a description of the packaging, a list of the tests, sampling time points for each of the tests, temperature and humidity conditions to be studied, expected duration of the stability program, and the proposed bracketing/matrixing protocol, if applicable. The specific analytical procedures should be referenced t the drug product specification section of the IND application or the USP, if possible.

A detailed data table that included the lot number, manufacturing site, the date of manufacture of the drug product, and the drug substance used to manufacture the lot should be provided. Each table should contain data from only one storage condition. Individual data points for each test should be reported. Representative chromatograms should be provided, if applicable.

A short description should be provided for each of the parameters being investigated in the stability program (i.e., stress, long-term, and accelerated) demonstrating that the appropriate controls and storage conditions are in place to ensure the quality of the product used in clinical trials. Tests unique to the stability program should be adequately defined.

For sterile products, the sponsor should consider the development of a container closure challenge test for future stability protocols. An appropriately designed test demonstrates that the container closure system can maintain the integrity of the microbial barrier during drug product shelf life. A discussion of how the selected test relates to the integrity of the container should be provided."[36]

The FDA recognizes that the amount of information about the stability profile of the product and its expiration date will be very limited in the early stage of clinical development. However, by Phase 3, it is expected that the stability profile is known and that an expiration date can be proposed.

5.2. Regulatory Expectations for the BLA/NDA Submission

By the time of filing of the dossier for market approval, the stability profile for both the API and the drug product, along with an expiry date, are expected to be justified. Table 74 presents the stability information to be included in the filing.

Table 74. Stability information to be included in the the BLA/NDA dossier for recombinant proteins and monoclonal antibodies[40, 41, 42, 172]

In-Process	Stability data supporting storage conditions should be provided.
API	Stability data are needed when the API is to be stored after manufacture, prior to formulation and final manufacturing
	A description of the storage conditions, study protocols and results supporting the stability of the drug substance should be submitted.
	Stability data from 3 lots representative of the manufacturing scale of production
	Minimum of 6 months stability data at time of submission
	Pilot-plant scale batches produced at a reduced scale of fermentation and purification may be submitted provided a commitment to place the first 3 manufacturing scale lots into the stability program after approval is made; also the storage conditions and containers must be the same (except a reduced container size can be used, if justified)
	Data from tests to monitor the biological activity and degradation products such as aggregated, deamidated, oxidized, and cleaved forms should be included, as appropriate.
	Data supporting any proposed storage of intermediates should also be provided.
	The postapproval stability protocol and stability commitment should be provided.
Intermediates	The manufacturer should identify any intermediates and general in-house data and process limits that assure their stability within the bounds of the developed process
	A description of the storage conditions, study protocols and results on the stability of intermediate fluids or formulated bulk under specified holding or shipping conditions, as appropriate, should be provided.
	While the use of pilot-plant scale data is permissible, the manufacturer should establish the suitability of such data using the manufacturing scale process
Drug Product	A description of the storage conditions, study protocols and results supporting the stability of the drug product should be provided.
	Stability data from 3 lots representative of the manufacturing scale of production

	Where possible, lots of drug product included in the stability program should be derived from different lots of API
	Minimum of 6 months stability data at time of submission
	Stability data from pilot-plant scale batches may be submitted provided a commitment to place the first 3 manufacturing scale lots into the stability program after approval is required
	For products administered through pumps or other such delivery devices, data on the stability of the drug product in the delivery system should be provided.
	The results of all tests used to monitor biological activity and the presence of degradation products such as aggregated, deamidated, oxidized, cleaved, etc. forms of the drug substance should also be included.
	The postapproval stability protocol and stability commitment should be provided.
Reconstituted Drug Product (for lyophilized products)	Stability of freeze-dried products after their reconstitution should be demonstrated for the conditions and the maximum storage period specified on containers, packages, and/or package inserts
	Stability data supporting the proposed shelf-life of the reconstituted drug product and for all labelled dilutions should be included.

It should be noted, that the amount of stability data required at time of submission for the API and drug product is different between chemically-synthesized drugs and biopharmaceuticals. Chemically-synthesized drugs require a minimum of 12 months real-time, real-condition stability data[173], while biopharmaceuticals require only a minimum of 6 months.[172]

It is the responsibility of the manufacturer to demonstrate the suitability of the lots selected for the stability program:

> "The quality of the batches ... placed into the stability program should be representative of the quality of the material used in preclinical and clinical studies and of the quality of the material to be made at manufacturing scale."[172]

Since only real-time, real-condition stability data can be used to determine the expiration date of the biopharmaceutical, it is most important for the manufacturer to understand what is actually in their ongoing stability program. This is a challenge during clinical development since numerous process changes are implemented to improve the manufacturing process.

5.3. Stability-Indicating Test Methods

Assay analytical procedures used in stability studies should be validated to be stability-indicating, unless scientifically justified:

> "A stability-indicting assay is a validated quantitative analytical procedure that can detect the changes with time in the pertinent properties of the drug substance and drug product. A stability-indicting assay accurately measures the active ingredients, without interference from degradation products, process impurities, excipients, or other potential impurities."[42]

There is no single stability-indicating assay or parameter that profiles the stability characteristics of a biopharmaceutical. The stability-indicating profile must include test methods that ensure that changes in the purity and potency of the product, indicative of degradation, will be detected. Other stability tests could include visual appearance of the product, pH determination, residual moisture of lyophilized products, sterility (or container/closure integrity). Stability of the excipients in the drug product also must be assured.

Product characterization provides the selection criteria for the release tests, and release tests provide the selection criteria for the stability tests. This is one of the reasons that the FDA recommends that stress testing be performed early during clinical development (preferably at Phase 2) in order to obtain a handle on which test methods could be stability-indicating.

While most biopharmaceutical manufacturers do not reveal their strategy for setting specifications, it is possible to get a glimpse at how it is done from CMC regulatory reviews.

Several examples from regulatory agency websites illustrate appropriate stability test method strategies for recombinant proteins and monoclonal antibodies. Table 75 (for an API) and Table 76 (for a drug product) compare test methods for release with those determined to be suitable for stability. Notice how the stability tests are a subset of the release test methods.

Table 75. Reported API release and stability testing strategy for the recombinant protein becaplermin (Regranex®)[174]

Test Parameters	Release	Stability
Appearance and Description	Appearance	Appearance
Identity (21 CFR 610.14)	Lys-C Peptide Mapping	

Purity (21 CFR 610.13)	SDS-PAGE Capillary electrophoresis SEC-HPLC RP-HPLC Endotoxin (LAL) Total DNA Yeast impurity ELISA Monosaccharide content	SDS-PAGE SEC-HPLC RP-HPLC
Potency (21 CFR 610.10)	Mitogenesis bioassay	Mitogenesis bioassay
Sterility (21 CFR 610.12)	Bioburden	Bioburden
Quantity	UV	UV
General Tests	pH	pH

Table 76. Reported drug product release and stability testing strategy for the monoclonal antibody infliximab (Remicade®)[71]

Test Parameters	Release	Stability
Appearance and Description	Appearance Color after reconstitution Visible particles after reconstitution Turbidity after reconstitution Reconstitution time	Appearance Color after reconstitution Visible particles after reconstitution Turbidity after reconstitution Reconstitution time
Identity (21 CFR 610.14)	Gel filtration-HPLC IEF Immunodifffusion assay	IEF
Purity (21 CFR 610.13)	Gel filtration- HPLC SDS-PAGE Endotoxin (LAL)	Gel filtration- HPLC SDS-PAGE
Potency (21 CFR 610.10)	Cell growth inhibition bioassay	Cell growth inhibition bioassay

Sterility (21 CFR 610.12)	Sterility	Sterility
Quantity	UV Uniformity of dosage units	UV
General Tests	pH Residual moisture	pH Residual moisture

Because there are now many recombinant proteins and monoclonal antibodies that have successfully been manufactured and reviewed by the regulatory agencies, stability programs are becoming fairly routine in terms of which test methods might be stability-indicating and what to expect in terms of long-term stability. However, the importance of the stability program increases as new technologies are applied to biopharmaceutical production. For example, for biopharmaceuticals from transgenic plants:

> "One concern regarding the lot-to-lot consistency of purified proteins made in plants is the limited experience with these products; consequently, there is little available data on long-term stability. It is possible that there will be unique proteases in plants that have deleterious effects on these products that have not been seen with prokaryotic, yeast or mammalian expression systems. This illustrates the important of rigorous real-time and accelerated stability studies with a new expression system."[72]

5.4. Setting an Expiration Date

Expiration dates are required for both the API and the drug product upon market approval. Expiration dates are not required during clinical development for either the API or the drug product, but it is required for the reconstituted product:

> "Section 11.6 ... An API expiry or retest date should be based on an evaluation of data derived from stability studies. Common practice is to use a retest date, not an expiration date... For new APIs, Section 11.6 does not normally apply in early stages of clinical trials."[22]

> "To assure that a drug product meets applicable standards of identity, strength, quality, and purity at the time of use, it shall bear an expiration date determined by appropriate stability testing ... New drug products for investigational use are exempt from the requirements of this section, provided that they meet

appropriate standards or specifications as demonstrated by stability studies during their use in clinical investigations. When new drug products for investigational use are to be reconstituted at the time of dispensing, their labelling shall bear expiration information for the reconstituted drug product."[175]

Some regional differences should be noted. The FDA prefers an expiration dating period for the API rather than a retest date. The FDA does not require expiration dates on clinical trial materials, which Canada and European countries do.[49]

It cannot be emphasized enough that primary data to support a requested storage period for a biopharmaceutical is based on long-term, real-time, real-condition stability studies.[168] Thus, the development of a proper long-term stability program becomes critical to the successful development of a commercial product.

Accelerated stability testing (typically storing the product at temperatures higher than that to be proposed for market conditions; freeze-thaw treatment) is useful for supporting the proposed expiration date and for assessing impact of future process changes, but accelerated stability testing data "do not substitute for real-time data for product approval and labelling."[29]

Stress stability testing (typically subjecting the product to intense light, heat, shear forces, oxidation conditions, etc.; hopefully, achieving at least a minimum of 10% change) is useful for validating the stability test methods and for assessing impact of accidental exposures of the product to conditions other than those proposed during transport, but stress testing data are only supportive of the proposed expiration date.

An 'expiration date' is the assigned shelf-life for either the API or the intermediate or the drug product. It is the period of time that the product can retain its specifications within established limits for safety, purity, and potency. An idea of the expiration dating periods that biopharmaceutical manufacturers are being granted by the FDA for market approval is provided in Table 77.

Table 77. Expiration dates listed in FDA approval letters for some biopharmaceuticals; listed on the FDA website, www.fda.gov/cber/products.htm

Biopharmaceutical Product	Expiration Dating Information in FDA Approval Letter
Tositumomab (Bexxar®)	API: redacted
	Drug Product: 36 months from the date of manufacture (defined as the date of final sterile filtration of the formulated drug product) when stored at 2-8°C

Alefacept (Amevive®)	API: 24 months from the date of manufacture when stored at -70°C ± 10°C
	Drug Product: 24 months from the date of manufacture (defined as the date of final sterile filtration of the formulated drug product) when stored at controlled room temperature (15-30°C, 59-86°F)
Interferon beta-1a (Rebif®)	API: 30 months when stored at -70°C
	Drug Product: 24 months from the date of manufacture (defined as the date of final sterile filtration of the formulated drug product) when stored at 2-8°C
Peginterferon alfa-2a (Pegasys®)	API: 18 months when stored at -70°C
	Intermediate (peginterferon drug substance): 18 months when stored at -70°C
	Drug Product: 12 months from the date of manufacture (defined as the date of final sterile filtration of the formulated drug product) when stored at 2-8°C

These assigned expiration dating periods are based upon the real-time/real-temperature data submitted for review in the BLA/NDA. If the manufacturer wants a 24 month shelf-life for its product, at least 24 months of stability data must be provided to the regulatory agency. This indicates the importance of stability planning, since the lots to be used to justify the shelf-life of the biopharmaceutical need to placed on stability sometimes even before Phase 3 clinical material is manufactured.

The FDA has its limits on what it will assign to the expiration dating period:

"With only a few exceptions, however, the shelf-life for existing products and potential future products will be within the range of 0.5 to 5 years."[172]

Clearly, for the typical manufacturer of a biopharmaceutical, a shelf-life of less than 18 months would be difficult to handle owing to the time it takes to get the manufactured products through Quality Assurance release and to the distribution centers. In reality, a shelf-life of 5 years is not commercially desirable since a manufacturer would like to have its inventory turned over to allow for continual manufacturing.

A major difference between chemically-synthesized drugs and biopharmaceuticals is in the of extrapolation shelf life. For chemically-synthesized drugs, limited extrapolation of the stability data beyond the available long-term data in order to extend the shelf life is possible. Under certain conditions, the extended shelf life can be up to 12 months.[173] However, for biopharmaceuticals, this is not allowed:

"Primary data to support a requested storage period ... should be based on long-term, real-time, real-condition stability studies."[172]

5.5. How Much Change is Acceptable?

For proteins, there are plenty of possible changes that can occur during storage:

- Chemical Instability

    Deamidation
    Oxidation
    Fragmentation
    Racemization
    Disulfide scrambling

- Physical Instability

    Denaturation
    Aggregation
    Precipitation
    Adsorption

The stability data from the ongoing stability program needs to be examined, the data trended and needs to be statistically analyzed in the BLA/NDA. The ICH Q1E document on 'Evaluation of Stability Data' provides guidance on how this analysis should be done:

"Where applicable, an appropriate statistical method should be employed to analyze the long-term primary stability data in an original application. The purpose of this analysis is to establish, withy a high degree of confidence, a retest period or shelf life during which a quantitative attribute will remain within acceptance criteria for all future batches manufactured, packaged, and stored under similar circumstances.

Regression analysis is considered an appropriate approach to evaluating the stability data for a quantitative attribute and establishing a retest period or shelf life. The nature of the relationship between an attribute and time will determine whether data should be transformed for linear regression analysis. Usually, the relationship can be represented by a linear or non-linear function on an arithmetic or logarithmic scale. Sometimes a non-linear regression can be expected to better reflect the true relationship.

An appropriate approach to retest period or shelf life estimation is to analyze a quantitative attribute by determining the earliest time at which the 95 percent confidence limit for the mean around the regression curve intersects the proposed acceptance criterion.

> For an attribute known to decrease with time, the lower one-sided 95 percent confidence limit should be compared to the acceptance criterion. For an attribute known to increase with time, the upper one sided 95 percent confidence limit should be compared to the criterion. For an attribute which can either increase or decrease, or whose direction of change is not known, two-sided 95 percent confidence limits should be calculated and compared to the upper and lower acceptance criteria."[176]

So when the statistical analysis is done, and change is measured, how should the change be evaluated? For biopharmaceuticals, there are no pre-assigned acceptable losses during shelf-life:

> "Maximum acceptable losses of activity, limits for physicochemical changes, or degradation during the proposed shelf-life have not been developed for individual types or groups of biotechnological/biological products but are considered on a case-by-case basis."[172]

This case-by-case approach is what makes it challenging for biopharmaceuticals. Again, what helps the manufacturer is the experience of the regulatory agencies with numerous other recombinant proteins and monoclonal antibodies that they have reviewed. Their body of information can sometimes allow the regulatory agencies to help the manufacturer put into perspective the change observed.

5.6. Applying the Minimum CMC Continuum to Stability

Sorting out which test methods could be stability-indicating early in the clinical development process can be invaluable. The risk to the manufacturer is that if the stability test methods are inappropriate, changes could be missed.

Knowing what is in the ongoing stability program is important. Many companies continue to test lots previously placed on stability even though the manufacturing process has significantly changed. To conserve resources, the stability program focus should be on the lots reflecting the current manufacturing process. However, it should be pointed out that as long as a lot of product is in clinical use, regardless of whether the process has changed or not, it should be supported by the ongoing stability program.

Figure 18 illustrates the minimum CMC continuum for stability.

Figure 18. Minimum CMC continuum for stability

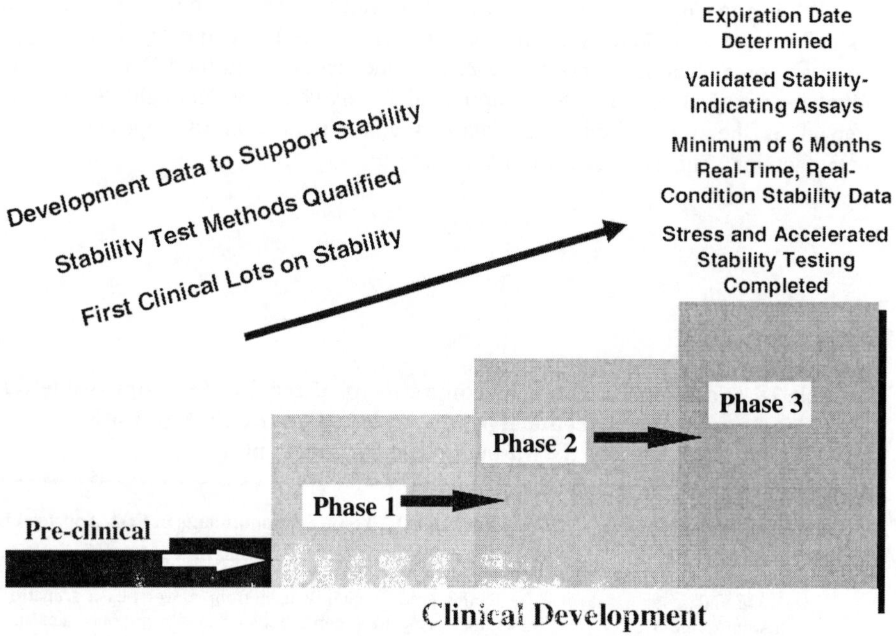

6. WHAT CAN GO WRONG

So much to do for a biopharmaceutical, from product characterization, to release test method selection and validation, to setting of specifications, to maintaining the ongoing stability program, to evaluation of the expiration date. Incomplete release and/or stability requirements in the BLA/NDA filing, major FDA 483 inspectional observations related to the QC release and stability testing of the product, product recalls, and misuse in the clinic are all part of the landscape of analysis problems that affect biopharmaceuticals.

6.1. Incomplete Release/Stability Testing Requirements in BLA/NDA Filing

While a manufacturer prepares what it considers to be a complete package to support its QC release and stability testing requirements, it is up to the regulatory agencies to concur. Sometimes this is not the case. One mechanism, that the FDA has used to allow biopharmaceutical products to be approved for market, even though QC testing issues remain, is the use of additional Phase 4 commitments in its approval letter to the manufacturer. This is illustrated in Table 78.

Table 78. Additional QC testing commitments listed in FDA approval letters for biopharmaceutical products; found on the FDA website www.fda.gov/cber/products.htm

Biopharmaceutical Product	Additional QC Testing Commitments in FDA Approval Letter
Laronidase (Aldurazyme®)	To submit product quality data from a worst-case scenario study at laboratory scale to support hold times for intermediates in the drug substance manufacture; results, including all tests required by the usual Certificate of Analysis, will be submitted to the FDA
Alefacept (Amevive®)	To conduct a validation study of container/closure integrity of bottles used to store bulk drug substance and to submit the protocol and final study report to the BLA
	We acknowledge your agreement to sample each filled bulk drug substance bottle for endotoxin and bioburden
Alemtuzumab (Campath®)	To evaluate data from the CMCL potency assay and the monomer content assay performed for lot release of bulk drug substance and drug product over the year following approval (or the next ten lots) and to tighten the lot release specifications appropriately with submission of a manufacturing supplement
Oprelvekin (Neumega®)	To reevaluate the stability specifications for the drug product
	To reevaluate the hold times for in-process intermediates
	To repeat the photostability study on lyophilized drug product
	To recalculate control limits for in-process yields
Drotrecogin alfa (Xigris®)	The drug substance and drug product specifications will be revised to include purity by SDS-PAGE analysis
	A specification for release of drug product to assure glycosylation remains consistent, will be developed

	Additional specificity studies regarding the APTT potency test will be submitted to the BLA
	A revised APTT potency test method utilizing a standard curve comprised of more than two data points will be submitted to the BLA
	An endotoxin assay with improved sensitivity for testing of the drug product will be developed and validated
	A drug product control sample, representing the commercial drug product composition, will be established and characterized; this control sample will be used for the RP-HPLC and SEC-HPLC identity/purity assay and the SEC-HPLC assay
	The IP-RP-HPLC, RP-HPLC and SE-HPLC purity methods for drug substance, as well as the RP-HPLC and SEC-HPLC methods for drug product, will be revised to require evaluation of the main peak for split peaks or shoulders in comparison to the relevant reference standard or control sample
Tenecteplase TNKase®	Institute action limits on bioburden for all stages of the manufacturing process

6.2. FDA 483 Inspectional Observations

Approximately 20% of all FDA 483 observations made by CDER in 2002 were related to the QC laboratory.[177] This percentage appears to be similar with CBER and Team Biologics inspections. Some of the QC laboratory problems noted for biopharmaceutical QC laboratories are presented in Table 79. It is surprising that out-of-specification (OOS) issues keep arising during the inspections, especially since the FDA has provided clear guidance on how these should be investigated.[178]

Table 79. FDA 483 observations related to QC laboratory operations[179, 180, 181]

Warning Letter Issued to Genetics Institute, December 23, 1999

Failure to maintain and follow an appropriate written testing program designed to assess the stability characteristics of drug products [21 CFR 211.166(a)], in that there was no data available to demonstrate that assays used in the stability testing of bulk Antihemophilic Factor Concentrate (Recombinant) (rAHF) are stability indicating.

Specifications have not been established for all in-process analyses performed on samples collected during the manufacture of Recombinant Human Interluken 11, Antihemophilic Factor Concentrate (Recombinant), and Coagulation Factor IX.

FDA Inspection of Genentech, Inc., July 24-August 3, 2001

Tenecteplase (TNKase) bulk drug substance lot #K11399 was placed on stability on 3/22/99. The specification for % single chain failed the acceptance criteria 4 times on stability study. A Non-Conformance Material

Report (NCMR) was not filed until approximately 13 months after the initial 9-month stability failure of this lot was experienced. This NCMR remains open (6 months after being filed). Furthermore, the firm failed to submit a Biological Product Report/Error & Accident to CBER to inform the Center of this stability failure.

Five lots of Rituxan exceeded the firm's % action limit during visual inspection. Investigation of the high percent of rejects for these lots revealed high levels of embedded foreign matter on the stoppers as the cause. The specific identity of the foreign matter has not been determined. No testing was performed to ensure the foreign material does not impact the product.

FDA Inspection of Eli Lilly and Company, October 15 – November 14, 2001

The most recent re-qualification records for 4 of the currently used stability chambers, were randomly selected for review. Personnel were found to have inaccurately recorded the dates they had actually performed various activities on documentation contained within each of the 4 records reviewed. All 4 re-qualification records, with the inconsistencies, had been signed off as reviewed and approved by the Quality Unit.

Concerning data handling and OOS, initial release test results for the Volatiles Free Potency testing for Lispro Human Insulin (recombinant) lot 523PAO failed internal specifications. The investigation did not determine that the original failing results were inaccurate. Despite the finding that the original failing results were determined to be accurate, the results ultimately reported by the firm was a passing result, which was the average of the original failing results with the passing results found during the investigational re-testing.

The firm's policy defines specific criteria to describe when failed drug product batches must be reported in an NDA annual report. Six recombinant human insulin lots and 1 recombinant human growth hormone lot had been listed as rejected within the last two annual report periods. None of these rejected lots were reported in annual reports to the FDA.

6.3. Biopharmaceutical Product Recalls

Regardless of the source of the drug product, chemically-synthesized or biopharmaceutical, occasionally there will be the need to recall a product that has been released and distributed into the marketplace. As shown in Table 80, the three primary reasons for biopharmaceutical product recall appear to be (1) loss of potency, (2) product handled outside of its approved temperature limits and (3) problems with the container/closure system or the drug delivery device.

The importance of maintaining the required temperature conditions for the drug product during handling and transport cannot be stressed enough. Typically, shipment validation is performed for the BLA/NDA filing, establishing the required shipping conditions and temperatures to be maintained. For biopharmaceuticals, many of which must be maintained at refrigerated conditions, this is referred to as 'cold chain management'.

**Table 80. Biopharmaceutical product recalls listed on the FDA websites;
www.fda.gov/cber/recalls.htm , www.fda.gov/opacom/Enforce.html**

Biopharmaceutical Product Manufacturer -- Date of Recall	Reason for Recall
Human Insulin Lispro Humalog® Eli Lilly – 03/03	Drug cartridges may be cracked or broken.
Denileukin diftitox Ontak® Seragen - 2/03	One box of the recalled product, which is to be maintained in the frozen state, was returned to the facility unfrozen, and inadvertently redistributed. The manufacturer disclosed that this condition may cause the product to become subpotent.
Human Insulin Novolin® Novo Nordisk - 12/02	Defective container; delivery system may deliver less than the expected amount of insulin.
Infliximab Remicade® Centocor - 11/02	In response to customer complaints associated with lack of vacuum in the vial, the manufacturer found that the recalled products exhibit a stoppering defect that may contribute to the high incidence of this occurrence.
Factor VIII Kogenate® Bayer - 3/02	Through ongoing studies, Bayer Corporation determined that the recalled lot may not meet minimum potency requirements when stored over two months at room temperature conditions. Product that has been consistently stored under refrigeration or at room temperature for less than two months remains acceptable.
Factor VIII Helixate FS® / Kogenate FS® Bayer - 1/02	Potency testing following nine weeks storage at room temperature found that the lots fell below the minimum potency specification. The lots continued to meet potency requirements when continuously refrigerated. The recall applies only to vials that were not stored under continuous refrigeration.
Human Insulin Lispro Humalog® Novartis – 1/02	Subpotency (last dose may deliver less insulin than expected due to breakage of pen internal parts).
Interferon alfa-2b Intron A® Schering - 10/01	Lots were found to be subpotent, after assaying as 87%-89%, at 30 days following reconstitution with sterile Bacteriostatic Water for Injection. Lot was withdrawn as a precautionary measure because of the potential that this batch would not meet the product quality specification for assay prior to its expiration date. All of the lots are within the product quality specification for assay when tested immediately upon reconstitution.

Factor VIII Helixate FS® / Kogenate FS® Bayer - 9/01	Potency testing following nine weeks storage at room temperature found that the lots fell below the minimum potency specification. The lots continued to meet potency requirements when continuously refrigerated. The recall applies only to vials that were not stored under continuous refrigeration.
Domase alfa Pulmozyme® Genentech - 4/01	Pulmozyme ampules were damaged during manufacturing. An investigation conducted by the firm disclosed that a small number of the ampules exhibited a tiny puncture resulting in leakage of the product.
Hepatitis B Vaccine Engerix-B Glaxo SKB - 2/01	There is a possibility that during shipment from Bindley Western Drug Company to their customers, the product was not maintained at the required temperature.
Domase alfa Pulmozyme® Genentech – 10/00	An unapproved re-filtration step was utilized to remove metal particulates that had been found in the concentrated bulk drug substance.
Factor VIII Kogenate® Bayer - 7/00	During the monitoring of product stability for potency, that test points within the 3 month period, at 25C storage, fell below 80% of the labeled potency value (76-78%). This causes concern that the effective dose may not be achieved if patients were to exercise the full 25C storage option. The apparent differences in potency are attributed to a reagent used during the initial testing. Potency continues to remain stable when the product is stored under refrigeration (2-8C).
Interferon alfa-2a Rogeron-A® Hoffman-La Roche - 6/00	The presence of an interferon degradant was found during the shelf life of the product.
Domase alfa Pulmozyme® Genentech – 4/00	Potential packaging defect which could result in ampule leakage

6.4. Misuse in the Clinic

During clinical trials, it is important that the biopharmaceutical product be maintained under storage conditions recommended by the manufacturer and that expired drug products not be used in the study. However, this is becoming a growing concern for the FDA especially as they are now auditing more clinical study programs. Manufacturers have to ensure that the clinical site is following its recommendations. If the FDA discovers a violation, the consequence could be the issuance of a Warning Letter to the clinical site or the possible invalidation of the clinical study. The following was recently described in a Warning Letter to a clinical investigator studying the use of a monoclonal antibody and two recombinant proteins:

> "You failed to ensure that three study drugs (Chimeric monoclonal antibody 14.18, GM-CSF, and IL-2) were stored at the required temperature to ensure the drugs maintained their specified characteristics. There is no documentation to show that the study drugs were in a refrigerator maintained at 2 to 8 degrees Centrigade."[182]

7. STRATEGIC CMC TIPS FOR BIOPHARMACEUTICAL ANALYSIS

The analysis of biopharmaceuticals requires consideration long-term planning. The regulatory agencies permit the start of clinical trials with minimal CMC in place; but they also require complete CMC by the time of BLA/NDA filing, as indicated in Figure 19. The challenge for a manufacturer is to have an action plan for filling in all of the CMC analysis information after Phase 1 has started.

Figure 19. Overall minimum CMC continuum for the analysis of biopharmaceuticals

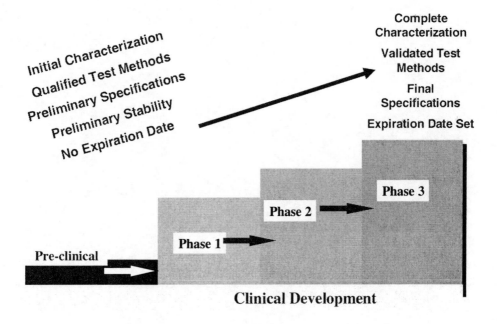

10

Managing Process Changes – Demonstrating Product Comparability

"Biological products and their manufacturing methods are quite complex and, in several instances, approval delays of several months or years have occurred due to the inability of the manufacturer to make a consistent product. Of note, in this regard, major upscaling and/or transfer of production to contract manufacturers frequently occurs late in biologic product development and applicants have difficulty showing that the product produced for market is comparable to the one used in clinical trials."

CBER explaining how manufacturing process changes occurring late in clinical development were impacting product approval timelines[183]

1. NOT AS EASY AS IT SEEMS!

During clinical development, manufacturers need the freedom to make changes (sometimes frequently) to make their processes sufficiently robust for commercialization. Even after market approval, changes may be necessary to increase the performance of the process, for line extensions, or to increase product label claims. But along with the freedom to make these changes, comes the responsibility of carefully assessing the impact on the product of such process changes.

Biopharmaceuticals, unlike chemically-synthesized drugs, are impacted in sometimes unpredictable, subtle ways by the manufacturing process, which can cause a change in either potency or immunogenicity. Therefore, whenever the process is changed, the product manufactured after the change must be compared to the product

287

manufactured before the change. The products cannot be different; but because biopharmaceuticals are complex products, they also most likely will not be identical. The biopharmaceutical products should be highly similar (Figure 8).

Figure 20. Possible outcomes of a product comparability study for a biopharmaceutical manufacturing process change

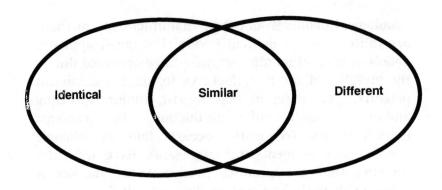

2. REGULATORY MANAGEMENT OF PROCESS CHANGES

Regulatory agencies readily accept that manufacturing process changes will occur during clinical development. In fact, they actually encourage such activities since they recognize that many biopharmaceutical manufacturing processes need considerable improvement and robustness built into them to ensure that consistent product could be manufactured at a later market approval stage.

"It is recognized that modifications to the method of preparation of the new drug substance and dosage form, and even changes in the dosage form itself, are likely as the investigation progresses."[16]

"Changes in the manufacturing scheme frequently occur during clinical development of a MAb."[29]

Process changes will occur, but the regulatory authorities require the manufacturer to demonstrate that the biopharmaceutical product is not impacted due to the change.

2.1. Pre-IND Stage

From a regulatory viewpoint, the easiest time to incorporate manufacturing process changes is prior to the start of human clinical studies (i.e., filing of the Phase 1 IND); but, for most biopharmaceutical companies, they are moving too fast to delay the start of clinical trials for this. So, typically, if a manufacturing process works to any extent, it is used to make the starting clinical trial materials, and process improvements are put off until later.

But there is one caution here: it is important to document any manufacturing process changes made between manufacturing of the pharm/tox materials for preclinical studies and the Phase 1 clinical trial materials. If changes have been made, the impact on the safety profile will need to be evaluated and included in the Phase 1 IND filing:

"Sponsors should describe any chemistry and manufacturing differences between the drug product proposed for clinical use and the drug product used in the animal toxicology trials that formed the basis for the sponsor's conclusion that it was safe to proceed with the clinical study. How these differences might affect the safety profile of the drug product should be discussed."[16]

2.2. IND Clinical Development Stages

Changes in the manufacturing process during the clinical development stage are to be communicated to the FDA through means of IND submissions and amendments. The regulatory agencies expect the manufacturer to have a well thought out plan for evaluating any proposed manufacturing process change:

"Sponsors should develop a plan for demonstrating that the products made by the old and new schemes are comparable, particularly when preclinical or clinical data developed prior to the production changes will be used to support further clinical trials and/or marketing applications. Similar considerations apply in the case of significant scale up in the manufacturing process (with or without modification of the general manufacturing scheme) implemented during or after completion of phase 3 trials."[29]

The regulatory agencies are also concerned that the manufacturer fully understands the difficulties in incorporating manufacturing process changes during the later stages of clinical development. As the clinical development program advances from Phase 1 towards Phase 3, a manufacturing process change has a greater potential impact on the existing efficacy and safety data that has already been generated. No one wins if a product approval is delayed because of an improperly evaluated manufacturing process change. For this reason, the FDA has two check points, one at the End-of-Phase 2 (EOP2) CMC meeting and the other at the pre-BLA/NDA CMC meeting, to discuss pending process changes with the manufacturer:

> "Examples of CMC issues that can be addressed at EOP2 meetings include ... major CMC changes, including site changes, anticipated from phase 2 through the proposed BLA/NDA, ramifications of such changes, and appropriateness of planned comparability and/or bridging studies, if applicable."[17]

> "Examples of CMC issues that could be addressed in pre-NDA or pre-BLA meetings include ... discussion of the relationship between the manufacturing, formulation and packaging of the drug product used in the phase 3 studies and the final drug product intended for marketing, and assurance that any comparability or bridging studies agreed upon at the EOP2 meeting have been appropriately completed."[17]

2.3. BLA/NDA Filing Stage

When submitting a BLA/NDA dossier to the FDA for market approval, it is important to include all required information. And this includes adequate information about manufacturing process changes and product comparability studies. The FDA has clearly stated what it expects in the filed BLA/NDA submission:

> "A discussion of any differences in formulation, manufacturing process, or site between the clinical trial materials and commercial production batches of drug substance and drug product should be submitted. If there are differences, a complete description of these differences should be included. If an investigational drug formulation was different form that of the to-be-marketed finished product, data to support comparability, bioequivalence and/or pharmacokinetic equivalence of the two formulations should be provided, if appropriate. If the manufacturing process and/or site was different, data from appropriate testing to assess the comparability of the investigational and commercial products should be provided."[40]

And the FDA has also clearly stated the consequences of what can happen if it doesn't find the required information in the BLA/NDA submission:

"A refusal to file decision may also be made in the absence of any of the following ... omission of data demonstrating equivalency to clinical trial product when significant changes in manufacturing processes or facilities have occurred ... failure to describe changes in the manufacturing process, from material used in clinical trials to commercial production lots."[31]

2.4. Post-Approval Market Stage

Upon market approval for a biopharmaceutical, the requirements of 21 CFR 601.12 (for a BLA)[184] and 21 CFR 314.70(g) (for a NDA)[185] on 'Changes to an Approved Application' become effective for managing future manufacturing process changes. These regulations permit a four-tier approach for informing FDA about all changes in the product, production process, quality controls, equipment, facilities, responsible personnel, or labelling, established in the approved license application:

1. Major Changes – those changes that have a substantial potential to have an adverse effect on the identity, strength, quality, purity, or potency of the product

    Requirement: Prior-Approval Supplement (PAS)

2. Moderate Changes – those changes that have a moderate potential to have an adverse effect on the identity, strength, quality, purity, or potency of the product

    Requirement: Change Being Effected Supplement, 30 days prior to distribution of the product (CBE30)

3. Moderate Changes – those changes that have a moderate potential to have an adverse effect on the identity, strength, quality, purity, or potency of the product

    Requirement: Change Being Effected Supplement (CBE)

4. Minor Changes – those changes that have a minimal potential to have an adverse effect on the identity, strength, quality, purity, or potency of the product

    Requirement: Annual Report (AR)

The FDA has provided a general roadmap of what manufacturing changes should be considered major (Table 81), moderate (Tables 82 and 83) or minor (Table 84). However, it should be pointed out, that frequently this general roadmap is not specific enough for the many changes contemplated by a biopharmaceutical manufacturer. In such cases, drawing upon the experience of a consultant familiar with assessing process changes, or even just asking the FDA reviewer for your specific product, can provide further guidance.

Table 81. Major process changes for a biopharmaceutical requiring a PAS submission to FDA[184, 185, 186]

Process changes including, but not limited to:

- Change in the source material or cell line
- Establishment of a new master cell bank or seed
- Extension of culture growth time leading to significant increase in number of cell doublings beyond validated parameters
- New or revised recovery procedures
- New or revised purification process, including a change in a column
- Changes in the virus or adventitious agent removal or inactivation method(s)
- A change in the chemistry or formulation of solutions used in processing
- A change in the sequence of processing steps or addition, deletion, or substitution of a process step
- Reprocessing of a product without a previously approved reprocessing protocol

Any change in manufacturing processes or analytical methods that:

- Results in change(s) of specification limits or modification(s) in potency, sensitivity, specificity, or purity
- Establishes a new analytical method
- Deletes a specification or an analytical method
- Eliminates tests from the stability protocol
- Alters the acceptance criteria of the stability protocol

Scale-up requiring a larger fermentor, bioreactor, and/or purification equipment (applies to production up to the final purified bulk)

Change in the composition or dosage form of the product or ancillary components (e.g., new or different excipients, carriers, or buffers)

New lot of, new source for, or different, in-house reference standard or reference panel (panel member) resulting in modification of reference specifications or an alternative test method

Extension of the expiration dating period and/or a change in storage temperature, container/closure composition, or other conditions, other than changes based on real time data in accordance with a stability protocol in the approved application

Change in the site(s) at which manufacturing, other than testing, is performed, addition of a new location, or contracting of a manufacturing step in the approved application, to be performed at a separate facility

Conversion of production and related area(s) from single to multiple product manufacturing area(s)

Addition of products to a multiple product manufacturing area if there are changes to the approved and validated cleaning and changeover procedures and additional containment requirements

Changes in the location (room, building, etc.) of steps in the production process which could affect contamination or cross contamination precautions

Changes which may affect product sterility assurance, such as changes in product or component sterilization method(s), or an addition, deletion, or substitution of steps in an aseptic processing operation

Changes requiring completion of an appropriate human study to demonstrate the equivalence of the identity, strength, quality, purity, or potency of the product as they may relate to the safety or effectiveness of the product

Table 82. Moderate process changes for a biopharmaceutical requiring a CBE30 submission to the FDA[184, 185, 186]

Addition of duplicated process chain or unit process, such as a fermentation process or duplicated purification columns, with no change in process parameters

Addition or reduction in number of pieces of equipment (e.g., centrifuges, filtration devices, blending vessels, columns. etc.) to achieve a change in purification scale not associated with a process change

An increase or decrease in production scale during finishing steps that involves new or different equipment

Replacement of equipment with that of similar, but not identical, design and operating principle that does not affect the process methodology or process operating parameters

Manufacture of an additional product in a previously approved multiple product manufacturing areas using the same equipment and/or personnel, if there have been no changes to the approved and validated cleaning and changeover procedures and there are no additional containment requirements

Change in the site of testing from one facility to another (e.g., from a contract lab to the applicant; from an existing contract lab to a new contract lab; from the applicant to a new contract lab)

Change in the structure of a legal entity that would require issuance of a new license(s), or change in name of the legal entity or location that would require reissuance of the license(s)

Table 83. Moderate process changes for a biopharmaceutical requiring a CBE submission to the FDA[184, 185, 186]

Addition of release tests and/or specifications or tightening of specifications for intermediates

Minor changes in fermentation batch size using the same equipment and resulting in no change in specifications of the bulk or final product

Table 84. Minor process changes for a biopharmaceutical requiring an AR submission to the FDA[184, 185, 186]

Increase in aseptic manufacturing scale for finished product without change in equipment, e.g., increased number of vials filled

Modifications in analytical procedures with no change in the basic test methodology or existing release specifications provided the change is supported by validation data

Change in harvesting and/or pooling procedures which does not affect the method of manufacture, recovery, storage conditions, sensitivity of detection of adventitious agents, or production scale

Replacement of an in-house reference standard or reference panel (or panel member) according to SOPs and specifications in an approved application

Tightening of specifications for existing reference standards to provide greater assurance of product purity and potency

Establishment of an alternate test method for reference standards, release panels, or product intermediates, except for release testing of intermediates licensed for further manufacture

Establishing of a new Working Cell Bank derived from a previously approved Master Cell Bank according to an SOP on file in the approved license application

Change in the storage conditions of in-process intermediates, which does not affect labelling, based on data from a stability protocol in an approved application

Change in shipping conditions (e.g., temperature, packaging, or custody) based on data derived from studies following a protocol in the approved application

A change in the stability test protocol to include more stringent parameters (e.g., additional assays or tightened specifications)

Addition of time points to the stability protocol

Change in the simple floor plan that does not affect production process or contamination precautions

Trend analyses of release specification testing results for bulk drug substances and drug products obtained since the last annual report

Any change made to comply with an official compendium that is consistent with FDA requirements

An extension of an expiration date based upon full shelf-life data obtained from a protocol approved in the application

A change within the container and closure system for solid dosage forms, based upon a showing of equivalency to the approved system under a protocol approved in the application or published in an official compendium

A change in the size of a container for a solid dosage form, without a change from one container and closure system to another

The addition or deletion of an alternate analytical method

Before the change can be implemented in the manufacture of a commercial product, even for those changes to be included in the annual report, the manufacturer must demonstrate that there is no product impact. Sometimes this demonstration may only require re-validation and/or lot release testing, but sometimes this demonstration may require a thorough product comparability study:

> "Before distributing a product made using a change, an applicant shall demonstrate through appropriate validation and/or other clinical and/or non-clinical laboratory studies, the lack of adverse effect of the change on the identity, strength, quality, purity, or potency of the product as they may relate to the safety or effectiveness of the product" [184, 185]

The manufacturer must also provide, in a submission to the FDA, the following documentation[184, 185].

1. A detailed description of the process change

2. A list of the product(s) involved

3. The manufacturing site(s) or areas(s) affected

4. A description of the methods used and studies performed to evaluate the effect of the change on the identity, strength, quality, purity, or potency of the product as they relate to the safety of the product

5. The data derived from such studies

6. Relevant validation protocols and data

7. A reference list of relevant standard operating procedures

It should be noted that EMEA has its own set of requirements for manufacturing changes after marketing authorization has been granted: Type I variations are minor process changes and Type II variations are major process changes, with corresponding requirements for information to be provided.[187, 188] Under revisions of Directive 2001/83/EC of the European Parliament, EMEA is moving towards a Type IA, Type IB and Type II system.[189] Table 85 presents some examples of manufacturing process changes approved under the EMEA variation system.

Table 85. Examples of some manufacturing process changes approved by EMEA for biopharmaceutical produts[190, 191]

Interferon beta-1b Betaferon® Schering AG	<u>Type I</u> • Tightening of shelf life moisture specification • Addition of a final filtration step • Replacement of Brain Heart Infusion (BHI) agar with a soya derived medium to be used in the manufacture of the working cell bank • Addition of a RP-HPLC test into the final product release testing and stability testing • For a second supplier of Betaferon: replacement of Brain Heart Infusion (BHI) agar with a soya derived medium to be used in the manufacture of the working cell bank • Addition of a suitable specification for the control of deamidated by-products in the final container product • Change in specification of excipients in the product • Change in storage conditions <u>Type II</u> • Solvent vial replaced with a pre-filled syringe
Rituximab Mabthera® Roche	<u>Type I</u> • Introduction of an additional manufacturer of the active substance (Genentech) with consequential changes an increase in batch size of the active substance and minor changes in the manufacturing process of the active substance and the medicinal product • Changes in the qualitative composition of intermediate packaging material • Addition of Roche as a filling site for Mabthera vials • Change in the seed-train of the manufacturing process for the API • Use of silicon oil for the vial stoppers • Extension of shelf-life <u>Type II</u> • Changes in the specifications for the sterile bulk

3. DEMONSTRATING PRODUCT COMPARABILITY

Whenever a manufacturing process change is considered, an assessment of the potential impact on the biopharmaceutical product must be made. This assessment requires understanding both the product and the process. During the early stages of clinical development, there will less analytical tools available for examining the product and less manufacturing process data to draw upon to make this assessment. Fortunately, demonstrating product comparability is less critical at this stage because definitive safety and efficacy have not yet been shown. But demonstrating product comparability will be more extensive and is more critical for those products in late stage clinical development or those products that have already been approved by the regulatory agencies.

3.1 Guidance Documents on Product Comparability

To assist biopharmaceutical manufacturers, both the FDA and EMEA have published guidance documents on how to determine product comparability:

- FDA Guidance Concerning Demonstration of Comparability of Human Biological Products, Including Therapeutic Biotechnology-Derived Products (1996)[28]

- EMEA Note for Guidance on Comparability of Medicinal Products Containing Biotechnology-Derived Proteins as Drug Substance (2001)[192]

- EMEA Note for Guidance on Comparability of Medicinal Products Containing Biotechnology-Derived Proteins as Drug Substance – Annex on Non-Clinical and Clinical Considerations (2002)[193]

It should be noted that ICH is also preparing a guidance document for product comparability.

3.2. A Three-Tiered Testing Hierarchy

The goal of a product comparability study is to demonstrate that the manufacturing process change does not affect safety, identity, purity or potency of the product. With biopharmaceuticals, it is difficult to predict the outcome of a comparability study. A minor or major alteration in a product characteristic can result in either no effect or a substantial effect in the safety and efficacy of the product. The same manufacturing change can have a variable effect on different products. It is most difficult to generalize the impact of a specific process change across all biopharmaceutical products.

Assessment of the impact of a process change on the biopharmaceutical is through a three-tiered testing hierarchy (Figure 21).

Figure 21. Three-tiered testing hierarchy for demonstrating product comparability

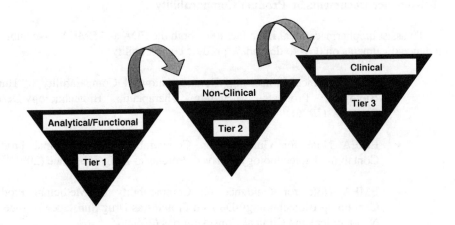

The first testing tier is the analytical/functional testing. This is laboratory testing of the products involving physicochemical, immunochemical and potency bioassays. If the manufacturer can demonstrate through this first tier of analytical/functional characterization that the products are comparable, then no further testing tiers have to be pursued. The manufacturer can make the argument to the regulatory agencies that the product has not been impacted by the change, thus inferring that the product from the changed manufacturing process has the same clinical efficacy and safety as the original product. However, if the first testing tier cannot demonstrate that the product is comparable, then the manufacturer must pursue the next testing tier(s).

The second testing tier is the non-clinical testing (or preclinical) testing. This is the use of animal models to assess pharmacokinetic parameters, pharmacodynamic activity, and/or toxicity endpoints. Many comparability studies include this second tier even when the analytical/functional testing demonstrates product comparability, because it can be complementary of the analytical/functional testing. For example, analysis of the pharmacokinetics profile often suggests biological events not reflected by the potency assays (e.g., bioavailability and effects on serum clearance). This second testing tier is especially needed in the following situations:

"Situations in which additional studies may be needed include those where the product has a narrow therapeutic range or where specific safety concerns are present, e.g., when the manufacturing process change raises concerns about possible toxic impurities or adventitious agents which cannot be assessed by analytical testing." [28]

"Depending on the quality of the data and the type of in vitro assays, the nature of the manufacturing change and the types of product differences observed or anticipated, a program of comparative testing (pharmacokinetics, etc.) in appropriate animal models may be considered in lieu of human clinical data when biochemical testing shows differences or cannot exclude significant differences in two products." [29]

"If there are concerns regarding safety, in vivo studies in one or more suitable animal models may be considered." [193]

The third testing tier is the clinical testing. This is the use of humans to study either simply the human pharmacokinetic profiles or the more extensive evaluation of human immunogenicty, safety and/or efficacy, due to a manufacturing process change. This third tier is hopefully to be avoided, because of the time involved in accruing a statistically sufficient number of patients into the study. This third testing tier is especially needed in the following situations:

"In cases where a manufacturing change(s) results in a product with structural and/or bioactivity differences, and/or differences in pharmacokinetic patterns, and those differences are meaningful with respect to potential impact on the product's safety, purity, or potency (efficacy), an additional clinical study(ies) usually may be needed to evaluate the product's safety and/or efficacy." [28]

"Additionally, when the analytical and other preclinical testing is not sufficiently sensitive or broad enough to detect such meaningful differences, additional clinical study(ies) may be needed." [28]

"In cases where animal studies may not be relevant, clinical pharmacology studies may be needed to show comparability." [28]

"Human pharmacology studies, generally, may be needed to evaluate changes which may affect product pharmacokinetics or pharmacodynamics, e.g., change in product formulation." [28]

"Comparative clinical evaluation of the products produced by different or scaled-up manufacturing schemes may be needed in certain situations: (a) product activity cannot be adequately characterized by analytical testing, (b) biochemical or biological testing show differences in the product, (c) animal testing reveals significant pharmacokinetic or other differences in the products and (d) the formulation of the product has been changed in a way that can affect its bioactivity." [29]

"Pharmacokinetic, safety and/or efficacy data may be required depending upon the nature and magnitude of the observed changes in the biochemical and/or biological properties of the product." [29]

"It should be recognized that in cases, where satisfactory comparability may not be demonstrable, a full preclinical and clinical data package will be required." [165]

"If a modification is detected during the comparability exercise, this either might imply a quality problem which might be resolved or it may indicate the need for further pre-clinical and/or clinical data." [193]

"If changes are likely to be present, but cannot be characterised, the efficacy and safety of the product need to be further characterised compared with the 'original' product" [193]

"... well characterized differences may provide a background for a rational approach with respect to the need for clinical studies ... In case the difference is non-major and there is no change in in vitro biological activity, comparable bioavailability data and/or pharmacodynamic studies may be sufficient. In case the difference is major, e.g., major changes in glycosylation pattern, or difference in in vitro activity, clinical equivalence studies will be necessary, unless otherwise justified." [193]

"Immunological studies are expected if a detailed and reliable physico-chemical characterisation is not possible due to the complexity of the molecule and an impact on immunogenicity by the changes to the manufacturing process cannot be excluded. In principle, major qualitative changes normally warrant immunogenicity studies pre-licensing." [193]

A manufacturer should be prepared to complete the appropriate testing tier(s) for the process change under consideration.

3.3. Designing the Comparability Study – Four Major Factors

Whenever change is introduced into the manufacturing process, product comparability must be assessed. The comparability study is the action plan that will be used by the manufacturer to demonstrate to the regulatory agencies that the product made after a process change is not impacted by that change. The following four major factors should be considered when designing the comparability study:

1. Clinical Development Stage for the Change

2. Where in the Process the Change is Introduced

3. Quality Criteria Considerations

4. Suitability of Available Analytical Methods

3.3.1. Factor 1: Clinical Development Stage for the Change

If the change is introduced during early clinical development stages, the level of understanding about the manufacturing process will be limited. In addition, the amount of analytical/functional testing available at the early clinical development stages will be limited. Many in-process controls will not have been established yet. Thus, the amount of testing available for the comparability study will be very limited.

At Phase 1, the comparability study will primarily consist of Certificate of Analysis (CofA) comparison of product manufactured before and after the change.

At Phase 2, the comparability study will consist of some product characterization, side-by-side testing and Certificate of Analysis (CofA) comparison of product manufactured before and after the change.

However, at the later stages of clinical development, and absolutely after market approval, the regulatory agencies expect at a minimum extensive analytical/functional testing in the comparability study.

3.3.2. Factor 2: Where in the Process the Change is Introduced

The location for the process change in the manufacturing process affects the design of the comparability study. For example, if the process change is to occur in the manufacturing process for the API, it is expected, that an analysis of in-process controls and process parameters, along with extensive analytical/functional testing of the product before and after the change, will be included in the comparability study. On the other hand, if the process change is to occur in the manufacturing process of the drug product, then maybe only comparative data on final release specifications and product stability data may be sufficient.

Major manufacturing process changes, as noted above, may automatically require the non-clinical and/or clinical testing in the comparability study. For example, if a formulation change is planned, human pharmacology studies are required.

It should also be noted that combining several minor manufacturing changes, may impact whether analytical/functional testing alone can be sufficient.

3.3.3. Factor 3: Quality Criteria Considerations

The complexity of the biopharmaceutical molecular entity is a major criterion in the design of the comparability study. Indeed, depending on the number of molecular variants for the molecule, it can sometimes be difficult to define precisely the product.

Comparability is more than 'meets specifications'. When designing the comparability study, the following 4 quality parameters should be given serious consideration: (1) product characterization studies, (2) variability of the manufacturing process, (3) release test data and specifications, and (4) stability data.

In the comparability assessment, it is important to be able to determine, that if a difference is observed, the source of the difference. Knowing the inherent variability of the manufacturing process allows the manufacturer to determine whether the product difference observed after the process change is within the normal variation experienced with the original manufacturing process or not. This is typically not known at the early clinical development stages. However, at later stage clinical development and after market approval, the variation due to the manufacturing process is determined through process validation studies and from the assessment of the conformance lots incorporating the manufacturing process change.

If the manufacturing process change results in a product with structural and/or bioactivity differences, non-clinical and/or clinical testing tiers will need to be considered.

3.3.4. Factor 4: Suitability of Available Analytical Methods

Biopharmaceuticals are complex and their inherent heterogeneity is sometimes difficult to determine. Extensive analytical/functional testing needs to be performed, but how relevant are the methods in detecting slight or discrete modifications of the product. It is the demonstration of the absence of these modifications that will justify the comparability of the product before and after the process change.

Therefore, in the comparability study, it is most important that the manufacturer justify the choice of analytical/functional tests to be performed. An evaluation is necessary of whether the analytical methods, at their existing level of validation, are suitable for detecting slight modifications possibly introduced by the change.

When analytical/functional testing is not sufficiently sensitive or broad enough to detect meaningful differences in the product, non-clinical and clinical testing tiers will need to be considered.

3.4. Regulatory Agencies Have Final Approval

It is most important to emphasize that it is the regulatory agencies, not the manufacturer, that ultimately approves a manufacturing process change:

> "If a manufacturer is able, in FDA's judgement, to demonstrate comparability, FDA may permit the manufacturer to implement the changes without conducting an additional clinical trial(s) to demonstrate efficacy."[28]

"FDA will determine if comparability data are sufficient to demonstrate that an additional clinical study(ies) is unnecessary."[28]

"FDA will determine the extent to which different types of comparability testing are necessary. For example, in some cases FDA may determine that no clinical study(ies) is necessary. In other instances, FDA may determine, on the basis of comparability data, that a clinical efficacy study(ies) is necessary."[28]

"FDA may determine that manufacturers of biological products, including therapeutic biotechnology-derived products regulated as biologics or drugs, may make manufacturing changes without conducting additional clinical efficacy studies if comparability test data demonstrate to FDA that the product after the manufacturing change is safe, pure, potent/effective."[28]

It is the goal of the manufacturer to provide adequate information to the regulatory agency so that they may understand the type of change made, the stage of production at which the change was made, and the product(s) affected. Nothing will speak stronger to the regulatory agencies than good science and solid data. Use good scientific judgement, and an honest critical evaluation, in interpreting the conclusions of the data. Don't try to hide a measurable product difference in the back of the regulatory report. Introduce the product difference early in the report and then provide the justification for why additional testing tiers are not necessary.

3.5 If in Doubt, Ask!

When the FDA published its guidance on product comparability, they emphasized strongly that a manufacturer should consider discussing their process changes in advance with them:

"In the interest of efficient review and approval of product applications, FDA encourages sponsors of unapproved applications or products under IND to consult with FDA regarding proposed manufacturing changes before implementing such changes prior to product approval."[28]

"FDA urges manufacturers to consult with FDA prior to implementing changes that may result in comparability testing, in order to avoid delay in the review of applications."[28]

"When changes in manufacturing occur during early clinical development, plans for evaluation of product comparability should be incorporated into product development strategies. Such plans should be discussed with CBER and, when appropriate, submitted to CBER for review."[29]

The FDA's desire is to prevent a manufacturer from doing all of the work demonstrating that a manufacturing process change has not impacted the biopharmaceutical product, only to find out later, that the regulatory agency has a different point of view of what is necessary to satisfy them of no product impact.

4. COMPARABILITY PROTOCOLS

A 'comparability protocol' is a well-defined, detailed, written plan for assessing the effect of a specific manufacturing process change. The comparability protocol describes specific manufacturing process change(s), and specifies the tests and studies that will be performed, including the analytical procedures that will be used, and acceptance criteria that will be achieved to demonstrate product comparability.[184, 185, 186]

The benefits of a comparability protocol are (1) that it impacts the timing of the approval for the process change and (2) after the FDA reviews the protocol, it is less likely to request additional information to support process changes. By all of the advanced planning and discussion with the FDA, the comparability protocol can permit a reduction in the regulatory review upon submission of the comparability report to the FDA (e.g., a PAS review ➔ a CBE30 review). This permits the manufacturer to implement the process change sooner after the comparability report has been submitted to the Agency.

However, it should be noted, that a comparability protocol does not reduce the amount of work that has to be done; it just distributes the work sooner in the regulatory review process. A comparability protocol does not negate the need for process validation. Also, it should be noted that a comparability protocol does not reduce the amount of work necessary to demonstrate product comparability.

If the acceptance criteria in the approved comparability protocol are met, then the manufacturer can distribute the product after submitting the comparability report to the Agency and within the timeframe of the reduced regulatory review. Manufacturers that include comparability protocols in their BLA/NDA for future process changes, typically find their comparability protocol acceptance mentioned in their approval letter from the FDA. Such was the case for the approval letter to Abbott Laboratories for the manufacture of adalimumab (Humira[TM]):

> "The comparability protocol for the manufacturing process in your license application is considered approved ... Although product made using this change may be distributed 30 days after FDA receives that supplement, continued use of the change will be subject to final approval of the supplement."[194]

What happens if the pre-defined acceptance criteria in the comparability protocol are not met after the product comparability study is completed? The manufacturer can elect either to not implement the change or to submit a PAS review after justifying why the missed acceptance criteria does not impact the safety and efficacy of the product.[18]

What has been the FDA's experience with comparability protocols? CBER management has communicated that manufacturers are getting their comparability protocols approved, but they also offered the following tips to the biopharmaceutical industry to help avoid some missteps[195]:

- Identify a specific process change or a group of related changes; do not bundle many process changes into the same comparability protocol

- It is helpful to develop pilot scale or full scale data to test the comparability protocol acceptance criteria prior to implementation; if these data are available, include them in the submission of the comparability protocol to the Agency.

- Above all do not alter your product specifications or your validation parameters, because once you do that you are automatically into a PAS review category

- Follow your approved comparability protocol; if you deviate, the Agency can decline the reduced reporting requirement.

For chemically-synthesized drugs, the FDA has issued its detailed guidance document on comparability protocols.[18]   But this document specifically excludes biopharmaceuticals. The FDA has recently issued a comparability protocol guidance document specifically for biopharmaceuticals.

5. CASE EXAMPLES OF BIOPHARMACEUTICAL COMPARABILITY

Successful product comparability doesn't just happen. There are a number of factors that can lead to its success:

- A well documented process development

- Comprehensive product characterization

- Good small scale models

- Good commercialization strategy

- Open dialogue with regulatory reviewers (early strategy discussions; transparent discussions; frequent updates of status, data and changes to plan)

- Partnership with regulatory reviewers who have access to proprietary information

5.1. Comparability Success Stories

Since the FDA and EMEA issued their guidances on product comparability, the biopharmaceutical industry has taken tremendous advantage of this approach in implementing manufacturing process changes. Various manufacturers have discussed their successes with product comparability studies (Table 86). These examples also demonstrate that other activities, other than product comparability studies, are needed to support the process changes (e.g., process validation, manufacture of consistency lots, etc.).

Table 86. Examples of process changes implemented after successful demonstration of biopharmaceutical comparability [128, 196, 197, 198]

Trastuzumab Monoclonal Antibody (Herceptin®), Genentech

Process Changes: Site change only, from South San Francisco to Vacaville, CA

Comparability demonstrated by release testing; additional characterization (peptide mapping, carbohydrate profiles, cation exchange HPLC); and n-process tests
Process validation (cell culture performance; recovery step yields; removal of residual DNA, host cell proteins, Protein A)
Consistency lots from Vacaville site compared to South San Francisco site (3 years of production)

Process Changes: Site change (California to Germany) + changes in the fermentation
(continuous feed vs bolus feed; removal of gentamicin in cell culture media),
in the harvest (use of centrifuge) and in the purification process (operation of Protein A
affinity column, additional concentration and diafiltration steps, introduction of a
nanofiltration step)

Feasibility at pilot scale indicated no product quality impact due to proposed changes
Process validation carried out for cell culture parameters, purification parameters and removal of impurities
Process re-validation carried out for lifetime of columns and virus removal
Comparability demonstrated by release testing; additional characterization (peptide mapping, oligosaccharide heterogeneity, charge heterogeneity); in-process tests (removal of DNA, host cell proteins, Protein A, endotoxin, bioburden); stability demonstrated for in-process samples and API (which included accelerated and stress testing).
Consistency lots after all process changes implemented

Palivizumab Monoclonal Antibody (Synagis®), MedImmune

Process Changes: Scale-up from 200L to 500L bioreactor size

Comparability demonstrated by extensive characterization and testing: all release tests, peptide mapping with mass spectrometry, binding activity, purity and microheterogeneity (oligosaccharide profile, deamidation, oxidation), residual impurity profiles and stability
Comparability also included pharmacokinetics

Process Change: Site change and scale-up to 10,000 L bioreactor size

Process validation carried out for cell culture parameters and cell line stability
Comparability demonstrated by extensive characterization and testing: all release tests, peptide mapping with mass spectrometry, binding activity, purity and microheterogeneity (oligosaccharide profile, deamidation, oxidation), residual impurity profiles and stability
Comparability also included pharmacokinetics
Consistency lots after all process changes implemented

Anakinra Interleukin-1 Receptor Antagonist (Kineret®), Amgen

Process Changes: Scale-up of API from 25 kg to 45 kg, adding a Working Cell Bank, some purification equipment changes and new recovery suite

New purification equipment validated
Pre-approval inspection (primarily because of new suite)
Comparability demonstrated by release testing; extensive characterization (primary, secondary and tertiary structural analysis); microheterogeneity; impurity profiles; 3-month stability data (including accelerated studies)
Consecutive conformance lots after process changes implemented

---

The European Public Assessment Report (EPAR) for Eli Lilly's Xigris® (recombinant human Activated Protein C) provides insight into how much change in the manufacturing process can be successfully managed during the clinical development stages by means of performing thorough product comparability studies:

"During the clinical development program several process modifications were implemented. Preclinical as well as Phase 1 and 2 clinical trial lots were manufactured during the first development process. Subsequent optimization studies led to the development of the initial commercial process which was used to manufacture lots for Phase 3 clinical trials. The latter process provided improved stability for the drug substance manufacturing solution. In changing from one to the other process a step was added to provide a greater level of viral safety assurance and certain animal-source raw materials were removed. Finally and during Phase 3 a slightly modified commercial process was introduced. This later one used a new master and working cell bank (cloned from the existing working cell bank) and the form of the drug substance was changed. In addition, the citrate concentration in the final elution step was reduced from 20 mM to 10 mM.

A summary of the range of assay values obtained for pilot-scale lots produced by the three mentioned manufacturing processes have been provided by the applicant. Although, variability for some parameters tended to be considerably greater for the first process, in general, the assay values observed for the first processed material were comparable to those observed using the later processes.

The applicant has performed extensive comparability studies between lots produced by the different processes used throughout the development of the drug substance production. The data provided show that the full-scale drug substance lots meet the proposed specifications, and they give a good indication of comparability between full-scale and pilot-scale lots. In addition to the specification assays, a diverse battery of additional characterization assays were performed to assess structural integrity and purity. It is also important to note that commercial scale lots were used in late phase clinical trials."[199]

One very wise decision on Eli Lilly's part was to ensure that after all of the manufacturing changes were implemented, that some of the product made by the commercial process was used in their Phase 3 clinical studies.

5.2. Comparability Surprises

Product comparability studies are not infallible. The following two case examples illustrate this point: interferon beta-1a and erythropoietin.

Avonex® is a recombinant interferon beta-1a derived from Chinese hamster ovary (CHO) cells. The pivotal clinical trials were carried out with a manufacturing process in Germany. For commercialization, the site for the manufacturing process was changed to the USA, and included a few process changes such as a new master cell bank and some improvements on product quality. Through extensive analytical/functional testing, and even some clinical pharmacokinetic testing, Biogen demonstrated that the products were comparable:

"BG9418 (to be marketed) has been extensively characterized and compared in side-by-side analyses with BG9015 (used in Phase 3). Biological, biochemical, and biophysical analyses have shown that the two molecules are comparable. Biological activities of each molecule are similar using several different assays, such as anti-viral, anti-proliferation, and enhancement of MHC class I expression. Peptide maps as determined by high pressure liquid chromatography of peptides derived by proteolysis of the two proteins are superimposable. Carbohydrate analysis revealed a similar pattern of major oligosaccharide forms on each protein. Finally, pharmacokinetic studies in humans using the two molecules revealed a pattern of clearance from the blood that was determined to be equivalent by rigorous statistical analyses."[200]

However, after the new product was in the market, the level of immunogenicity significantly decreased. About 24% of patients treated with the former product developed neutralizing antibodies, while only about 5% of patients treated with the new product developed neutralizing antibodies.[201] The process changes in the new manufacturing process are suspected to be the most likely cause of the reduced immunogenicity.[202]

Eprex® is a recombinant erythropoietin (EPO) derived from CHO cells. After market approval in Europe, the formulation of the drug product was changed (i.e., Human Albumin replaced with non-protein excipients) along with a few other minor manufacturing changes. Through comparability testing, Johnson and Johnson demonstrated that the products were comparable. However, after the new formulated product was introduced into the market, the level of immunogenicity significantly increased in the patients receiving the drug product, resulting in over a hundred life threatening cases of pure red cell aplasia due to neutralizing antibodies to EPO.[202] This unpredicted serious adverse event has caused the regulatory agencies to reconsider their confidence in analytical/functional test results.[203]

5.3. Not Comparable

Most manufacturing changes do not impact biopharmaceutical. But at times, some process changes can significantly impact the biopharmaceutical. While manufacturers are willing to talk about their successes with product comparability studies, they are reluctant to discuss their failures. However, the FDA has been more willing to discuss the failures (Table 87).

Table 87. Process changes that impacted the biopharmaceutical[102, 204, 205]

Process Change	Impact on Biopharmaceutical Product
Change in process water for fermentation	
Changed from potable to purified	Increased growth rate of host cells which led to a decreased specific activity for a product intermediate
Changed from purified to deionized	Amino acid substitutions in expressed product which caused the product not to meet release specifications
Change in Master CellBank (recloned), change in site of manufacture, with some changes made to fermentation medium and process configuration for a monoclonal antibody (This was done after Phase 3, but before market approval)	The nonoclonal antibody product had an amino acid substitution and altered glycosylation after the process changes Result: Product was not allowed to proceed to market

| Change in ultrafiltration process to target a lower SDS concentration, but still within existing specifications, for a recombinant IL-2 (This was done during Phase 3) | Lowering SDS levels, though within specifications, resulted in increased product purity, but also caused increased aggregation, changes in biodistribution and changes in animal safety and efficacy models

Specification for SDS concentration had to be tightened to 160-195 ug SDS/mg IL-2 |
|---|---|
| Change in site of manufacturer | |
| Same manufacturer | Purity declined, due to a procedure changes (e.g., pH during extraction, purification column operation) |
| Different manufacturer | Comparable recombinant protein was produced, but pK studies showed differences, due to small differences in glycosylaiton |
| Different manufacturer | Analytical comparability for the monoclonal antibody was obtained; but functional comparability was not (decrease in affinity in binding assay), which resulted in impaired clinical efficacy |

Finding that the product has been altered by a manufacturing process change can be a major setback for a biopharmaceutical company. For example, when Genentech took over manufacturing of Raptiva™ monoclonal antibody from Xoma for the Phase 3 clinical supplies, Genentech scaled-up the process and modified the process for its manufacturing operation. Extensive analytical/functional testing indicated that the product was comparable after the manufacturing change. However, pharmacokinetic studies showed a slightly higher serum concentration with the new product. The FDA was unhappy with the lack of bioequivalence, and the companies were forced to delay the BLA submission for 18 months.[206]

Biopharmaceutical CMC Outsourcing

> *"Manufacturer* means any legal person or entity engaged in the manufacture of a product subject to license under the act; *Manufacturer* also includes any legal person or entity who is an applicant for a license where the applicant assumes responsibility for compliance with the applicable product and establishment standards."
>
> FDA regulation allowing biopharmaceutical companies to be the
> applicant for a license even when they contract out
> their manufacturing and quality activities[207]

1. REGULATORY EXPECTATIONS FOR CONTRACTED WORK

When resources are abundant, a biopharmaceutical company can dream of becoming FIPCO (a fully integrated pharmaceutical company); however, when resources are limited, which is typical of most biopharmaceutical companies, serious consideration is given to outsourcing select manufacturing or quality activities.

1.1. Why Outsource?

The development of complex and highly specialized technology and equipment for the manufacture and testing of biopharmaceuticals has fostered the emergence of many companies that are capable of performing only limited aspects of manufacturing processes and/or release or stability tests. Consequently, many biopharmaceutical companies, to facilitate product development, are interested in contracting parts or all of their manufacturing and/or testing to these contractors.

There are benefits for a biopharmaceutical company to outsource some or all of its operations. Firstly, the internal personnel are kept to a minimum, netting benefits in permanent headcount costs. Secondly, the capital costs of designing, building and maintaining a facility is minimized. Thirdly, and probably most importantly, the use of experienced third parties could eliminate costly oversights and mistakes typically made by people with less experience or competence.

Most biopharmaceutical companies utilize some type of CMC outsourcing. The outsourcing could be as simple as contracting out cell bank characterization virus testing or the viral safety evaluation testing. The outsourcing could be more extensive such as contracting out the cell culture operation or the drug product filling/vialing operation. Some biopharmaceutical companies are 'virtual', with almost all activities contracted out.

1.2. Written Quality Agreements Required

Whether the manufacturing or testing activity is performed by the biopharmaceutical company at its own facilities or contracted out, the same cGMP regulatory expectations apply. But the regulatory agencies hold the biopharmaceutical companies accountable for the compliance at the contractor:

> "Companies should evaluate any contractors (including laboratories) to ensure GMP compliance of the specific operations occurring at the contractor sites."[22]

> "The quality control unit shall be responsible for approving or rejecting drug products manufactured, processed, packed, or held under contract by another company."[208]

The biopharmaceutical company needs to satisfy the regulatory agencies that it can manage the compliance of its contactor. Both the FDA and EMEA require that this compliance management plan be in a written contract:

> "Because the applicant assumes responsibility for compliance at the contract site ... the responsibilities of each party should be described in written agreements between the license applicant and the contract manufacturer."[209]

> "There shall be a written contract covering the manufacture and/or analysis arranged under contract and any technical arrangements made in connection with it."[210]

> "There should be a written and approved contract or formal agreement between a company and its contractors that defines in detail the GMP responsibilities, including the quality measures, of each party."[22]

Remember their philosophy, if it isn't written, it doesn't exist.

The 'intercompany quality agreement', the IQA, (sometimes called the 'quality supply agreement', the QSA) provides the written, defined compliance management plan. While the IQA could be included in the manufacturing supply contract (sometimes called the 'business contract'), it is now being treated more and more as a separate contract. Keeping it separate from the legal and business arrangements, offers two keep advantages: (1) flexibility in updating or changing , since only the two Quality Units need to agree and (2) lawyers are not required (the IQA can be brought in under the GMP document control system)..

1.3. Regulatory Requirements During Clinical Development

For the Phase 1 and Phase 2 IND submissions, the FDA primarily wants to know the identity and address for the manufacturer of both the API and the drug product.[16, 36] If this manufacturing is contracted out, the contractor would be the one identified.

For the Phase 3 IND submission, the FDA wants to have a full list of all contracted activities:

> "A list of all firms associated with the manufacturing and controls of the drug substance should be provided, including contract laboratories for quality control and stability testing."[36]

> "A listing of all firms associated with the manufacturing and controls of the drug product should be submitted, including the contractors for stability studies, packaging, labelling and quality control release testing."[36]

There is no mention of needing an IQA during clinical development. However, it is never too early to start the process:

> "You can look at outsourced drug development as a courtship and marriage. Phase 1 clinical trial material is produced during the courtship between two companies, Phase 2 during their romance, and by Phase 3 the companies are more or less 'married' to each other. After the pivotal lots are made, that marriage has produced children (precious clinical lots for submission to FDA), and it is too late to decide that things are not working out."[211]

1.4. Regulatory Requirements for the BLA/NDA Submission

During the pre-BLA/NDA CMC meeting, the biopharmaceutical company will need to assure the FDA that their contractor(s) is also ready for a pre-approval inspection at the time the dossier is submitted[17]

For the market approval dossier submission, the FDA expectation is not only that there is a complete list of all contactors provided, but also that the responsibilities of each contractor is defined:

> "The application should include the name(s), address(es), FDA registration number, and other pertinent organizational information for each manufacturer performing any potion of the manufacture or testing operations for the drug substance. This may include contractors or company subsidiaries serving as contractors, or other locations/sites owned and operated by the applicant. A brief description of the operations performed at each location, the responsibilities conferred upon each party by the applicant and a description of how the applicant will ensure that each party fulfills their responsibilities should be submitted." [39]

> "The name(s) and address(es) of all manufacturers involved in the manufacture and testing of the drug product including contractors, and a description of the responsibility(ies) of each should be submitted." [39]

> "The name, address, and responsibilities of each manufacturer, including contractors, and each proposed production site or facility involved in manufacturing and testing should be provided." [42]

1.5. Regulatory Requirements Post-Market Approval

The regulatory agencies have a clear idea of the openness of information flow and the level of cooperation that it expects between the contractor and the biopharmaceutical company:

> "The applicant should have access to floor plans, equipment validation, and other production information from the contract site necessary to assure safety, purity and potency of the product. The applicant should be fully informed of all deviations, complaints, and adverse events, as well as the results of all tests and investigations regarding or possibly impacting the product. The contract manufacturer should also share with the applicant all important proposed changes in production and facilities (including introduction of new products)." [209]

> "A contract should permit a company to audit its contractor facilities for compliance with GMP." [22]

> "Changes in the process, equipment, test methods, specifications, or other contractual requirements should not be made unless the contract giver is informed and approves the changes." [22]

2. DEVELOPING THE INTERCOMPANY QUALITY AGREEMENT

The biopharmaceutical company that needs to outsource some aspect of its manufacturing/testing and the contract manufacturer that offers such services, do not naturally have shared objectives. Assuming that both parties will be guided only by good will, is a dangerous position in this business. While both parties may agree to work together at the start, conflict often arises when business needs and incentives of the two parties diverge. The IQA, hopefully, can minimize some of the conflict.

2.1. Two Viewpoints

In many areas of the pharmaceutical industry, it is standard practice to establish separate, independent contract manufacturers not only to ensure an adequate supply of drug product but also to ensure sustained leverage through competition of multiple suppliers. In contrast, most biopharmaceuticals are single-sourced. The concept of redundancy is rare in biopharmaceutical manufacturing, except when adequate capacity can only be accomplished by more than one manufacturing site. Most biopharmaceutical companies cannot offer the duplication of effort and cost associated with establishing an additional manufacturer. Also, comparability of the biopharmaceutical produced at the additional site of manufacture must be proven and there is always the risk, that after all of the costs are expended to bring the new site on-line, that the products may not be comparable.

2.1.1. The Biopharmaceutical Company Seeking to Outsource

When a biopharmaceutical company decides to initiate clinical studies, it faces the challenge of obtaining these clinical supplies. The company would like to have their product manufactured in the most rapid and cost-efficient manner. The contractor, on the other hand, based on its past experience with other biopharmaceutical clients, has a certain degree of skepticism about how well the proposed manufacturing process is really defined and under control. This frustrates the biopharmaceutical company when the timelines are not within their expectations and the contractor proposes more resources (hence, a higher cost) to do a satisfactory technology transfer.

As the biopharmaceutical product moves through clinical development, there will be the need to manufacture larger quantities to keep the clinical studies supplied. The biopharmaceutical company would like timely access to the use of the contract facility whenever such a need arises. The company then finds it frustrating when the contract manufacturer responds that their facility is already outsourced to other clients for some period of time.

The biopharmaceutical company may initially consider moving the manufacturing process to another contractor, but then realizes the cost of such a technology transfer, and that the moved manufacturing process could impact the safety and efficacy of the biopharmaceutical product. Clinical studies are relatively lengthy and expensive, and no company wants to unnecessarily risk them. Therefore, the biopharmaceutical company becomes frustrated because they cannot easily 'take their business elsewhere'.

2.1.2. The Contact Manufacturer Offering Outsourcing

The contract manufacturer is in business to make a profit. They have services to provide, but they must pay for their overhead whether they have clients or not.

Unlike many well-established pharmaceutical companies, most biopharmaceutical companies are not large, profitable, stable organizations with significant financial resources. Contract manufacturers recognize that the probability of given biopharmaceutical product reaching commercialization is relatively low; however, each of their clients is highly confident that their product will succeed.

Contract manufacturers generally have more experience than their clients. The trouble with being an 'expert' is that they have the experience in knowing what it really will cost (both in terms of time and expense) to bring the biopharmaceutical product into their facility, but the clients are frequently unwilling or unable to fund the appropriate, controlled, and scientifically-based way of implementing technology transfer or process development.

The contract manufacturer is typically unwilling to divulge too many process details to a client. After all, their business is dependent upon being able to do a service that others cannot readily provide. There will always be the fear that they are 'educating the client' so that the client can take the process elsewhere.

While the contract manufacturer would very much like to have a guaranteed long-term contract with the biopharmaceutical company, this probably is the last thing on the mind of a financially-struggling biopharmaceutical company.

2.1.3. Maximum Leverage

The biopharmaceutical company has minimum dependency and, therefore, maximum leverage with a contract manufacturer before a manufacturing supply contract is signed. To achieve the signing of the IQA in a timely manner, will require some negotiation and even compromising by both parties.

Once contract manufacturing has begun, and definitely after the manufactured biopharmaceutical is used in a company's clinical trials, the leverage of the biopharmaceutical company over its contractor is significantly reduced. It becomes most difficult to negotiate an IQA at this point. Most of the time, the biopharmaceutical company has to live with what the contractor is willing to do. Fortunately, there are many reputable biopharmaceutical contractors that have the experience and wisdom to carry out their activities under proper cGMPs, and they understand what the regulatory agencies expect them to provide to the biopharmaceutical companies. However, it is a major risk to the biopharmaceutical company if they have not chosen such a contractor.

2.2. Contents of an IQA

A well written IQA's primary purpose is to cover the cGMP-related compliance issues. Since the document will be shown to regulatory inspectors, it should not contain business details or costs or quantities of product under contract to be supplied. The IQA must also clearly describe the quality and compliance roles and responsibilities between the biopharmaceutical company and its contractor. The IQA must demonstrate how the companies will ensure that the safety, purity, and potency of the biopharmaceutical will not be compromised as a result of the contract arrangement. Table 88 presents what the FDA expects to see in the signed, written agreement.

Table 88. Requirements to be included in the written quality agreement[209]

Identification of the contract manufacturer and the locations to be used for manufacture of the applicant's product

A description of the responsibilities of each participant, including the quality assurance function of the contractor and the supervision and control exercised by the license applicant

A description of the product shipped to the contract facility

Information describing the manner and conditions of shipment of product to and from the contract facility

A description of the operations(s) to be performed at the contract facility

The standard operating procedures to be used applicable to the contract arrangement, including procedures used to segregate manufacturing of different products

A commitment from the contract facility to inform the license applicant of all proposed changes in manufacture and facilities prior to implementation, including introduction of additional marketed products and clinical material processing operations

A commitment form the contract facility to fully inform the license applicant of all errors and deviations in manufacturing methods and test results, as well as adverse events, for the affected product

A description of how and when the contract facility will be periodically assessed by the license applicant for compliance with applicable product and establishment standards and cGMP

Table 89 presents an IQA template that I have used to initiate discussion between a biopharmaceutical company and its contractor.

Table 89. Example of an IQA table of contents

3. STRATEGIC CMC TIPS FOR OUTSOURCING

Not every expectation can be penned into an IQA. Therefore, the choice of the contract manufacturer can mean the difference between achieving the milestones and goals for the biopharmaceutical company or not. Choose wisely.

According to the FDA, one of the biggest problems that companies have when using a contract manufacturer, is open communication.[212] Relationships have to be nurtured. It is the wise head of Quality at the biopharmaceutical company that views the contractor's Quality Unit as a partner.

I have found that many contractors already have a basic IQA document available for the biopharmaceutical company to review, but the client needs to initiate the request:

> "While Quality Agreements have become a significant part of our business, it is limited to a specific segment of our clientele, Big Pharma. Our virtual or emerging biotech customers typically do not ask for Quality Agreements, but rather focus on the Supply Agreement alone."[212]

12

Concluding Thoughts on Biopharmaceutical CMC Regulatory Compliance

> "**Pharmaceutical manufacturing is evolving from an art form to one that is now science and engineering based. Effectively using this knowledge in regulatory decisions in establishing specifications and evaluating manufacturing processes can substantially improve the efficiency of both manufacturing and regulatory processes.**"
>
> FDA summary of the need for a risk based approach to cGMPs in the 21st century[214]

Biopharmaceutical CMC regulatory compliance is not static. Just during this past year alone, while I was writing the book, the FDA announced its decision to move many biopharmaceutical products from CBER review over to CDER review, they announced a major initiative on cGMPs for the 21st century, they issued many CMC guidance documents on both format (e.g., using the CTD for the BLA/NDA drug product section) and content (e.g., biopharmaceuticals from bioengineered plants), and they approved for marketing 7 new biopharmaceutical products.

Therefore, the need to stay current and constantly educate oneself to changing biopharmaceutical CMC regulations and guidance is most important. As I meet with various biopharmaceutical companies, I cringe when I hear the statements "well, we did it that way 5 years ago" or "it worked well for us the last time we did it". This is a dangerous position to take especially in such an evolving area like biopharmaceutical CMC regulatory compliance.

1. MOST HELPFUL WEBSITES FOR BIOPHARMACEUTICALS

As I mentioned at the beginning of the book, the ready access provided by the internet has opened the door for obtaining a wealth of important and practical information and guidance on biopharmaceutical CMC regulatory compliance. Take advantage of it.

Table 90 presents my top ten favourite websites.

Table 90. Website resources for biopharmaceutical CMC regulatory compliance

FDA Center for Biologics Evaluation and Research (CBER)	http://www.fda.gov/cber/
FDA Center for Drugs Evaluation and Research (CDER)	http://www.fda.gov/cder/
Office of Regulatory Affairs (ORA)	http://www.fda.gov/ora/
The European Agency for the Evaluation of Medicinal Products (EMEA)	http://www.emea.eu.int/
International Conference on Harmonization (ICH)	http://www.ich.org/
International Association for Biologicals (IABS)	http://www.iabs.org/
PDA	http://www.pda.org/
Regulatory Affairs Professional Society (RAPS)	http://www.raps.org/
European Directorate for the Quality of Medicine (EDQM)	http://www.pheur.org/
European Commission	http://pharmacos.eudra.org/F2/home.html

2. WEBSITE RESOURCES FROM FDA

Both CBER and CDER websites provide much biopharmaceutical CMC regulatory compliance resources. Both websites provide information on the products that have been approved, a list of guidance documents, and extensive safety information (e.g., Warning Letters issued, recalls, etc.). An extremely useful feature available on both the CBER and the CDER websites is the automatic "What's New" email notification. Whenever a new page or new information is added to the website, it automatically sends you an email notifying you of its availability. Some of the CMC regulatory guidances available on the CBER website are presented in Table 91. Some of the CMC regulatory guidances available on the CDER website are presented in Table 92.

Table 91. CMC regulatory guidances available on the CBER website; www.fda.gov/cber/guidelines.htm

Draft Guidance for Industry: Continuous Marketing Applications: Pilot 1 – Reviewable Units for Fast Track Products Under PDUFA - 6/12/2003

Draft Guidance for Industry: Continuous Marketing Applications: Pilot 2 – Scientific Feedback and Interactions During Development of Fast Track Products Under PDUFA - 6/12/2003

Draft Guidance for Industry: Developing Medical Imaging Drug and Biological Products - 5/19/2003

Guidance for Industry: Source Animal, Product, Preclinical, and Clinical Issues Concerning the Use of Xenotransplantation Products in Humans - 4/3/2003

International Conference on Harmonisation (ICH); Guidance for Industry: M2 eCTD: Electronic Common Technical Document Specification - 4/1/2003

Draft Guidance for Industry; Comparability Protocols - Chemistry, Manufacturing, and Controls Information - 2/20/2003

International Conference on Harmonisation (ICH); Guidance for Industry: Q3A Impurities in New Drug Substances - 2/11/2003

Draft Guidance for Industry: Drug Product: Chemistry, Manufacturing, and Controls Information - 1/28/2003

International Conference on Harmonisation (ICH); Guidance for Industry; Q1D Bracketing and Matrixing Designs for Stability Testing of New Drug Substances and Products - 1/15/2003

International Conference on Harmonisation (ICH); Draft Guidance on the M4 Common Technical Document-- Quality: Questions and Answers/Location Issues - 12/27/2002

Draft Guidance for Industry: Nonclinical Studies for Development of Pharmaceutical Excipients - 10/1/2002

Draft Guidance for Industry: Drugs, Biologics, and Medical Devices Derived from Bioengineered Plants for Use in Humans and Animals - 9/6/2002

International Conference on Harmonisation (ICH); Draft Consensus Guideline; Evaluation of Stability Data - 6/13/2002

International Conference on Harmonisation (ICH); Draft Consensus Guideline; Stability Data Package for Registration in Climatic Zones III and IV- 6/13/2002

Guidance for Industry: Container Closure Systems for Packaging Human Drugs and Biologics; Questions and Answers - 5/13/2002

Guidance for Industry: Providing Regulatory Submissions to CBER in Electronic Format - Investigational New Drug Applications (INDs) - 3/26/2002

Guidance for Industry: General Principles of Software Validation; Final Guidance for Industry and FDA Staff - 1/11/2002

International Conference on Harmonisation (ICH); Guidance on M4 Common Technical Document - 10/16/2001

International Conference on Harmonisation (ICH) Guidance; Q7A Good Manufacturing Practice Guide for Active Pharmaceutical Ingredients - 9/25/2001

Draft Guidance for Industry: Submitting Marketing Applications According to the ICH-CTD Format - General Considerations - 9/5/2001

Draft Guidance for Industry: Submitting Type V Drug Master Files to the Center for Biologics Evaluation and Research - 8/22/2001

Draft Guidance for Industry: Biological Product Deviation Reporting for Licensed Manufacturers of Biological Products Other than Blood and Blood Components - 8/10/2001

Guidance for Industry: IND Meetings for Human Drugs and Biologics; Chemistry, Manufacturing and Controls Information - 5/25/2001

Guidance for Industry: Monoclonal Antibodies Used as Reagents in Drug Manufacturing - 3/29/2001

International Conference on Harmonisation; Guidance on Q6A Specifications: Test Procedures and Acceptance Criteria for New Drug Substances and New Drug Products: Chemical Substances - 12/29/2000

Guidance for Industry: Supplemental Guidance on Testing for Replication Competent Retrovirus in Retroviral Vector Based Gene Therapy Products and During Follow-up of Patients in Clinical Trials Using Retroviral Vectors - 10/18/2000

Guidance for Industry: Q & A Content and Format of INDs for Phase 1 Studies of Drugs, Including Well-Characterized, Therapeutic, Biotechnology-Derived Products - 10/3/2000

Draft Guidance for Industry: Analytical Procedures and Methods Validation - Chemistry, Manufacturing, and Controls Documentation - 8/30/2000

International Conference on Harmonisation (ICH) Draft Revised Guidance on Impurities in New Drug Products - 7/19/2000

International Conference on Harmonisation; Draft Revised Guidance on Q1A(R) Stability Testing of New Drug Substances and Products - 4/21/2000

Guidance for Industry: Formal Meetings With Sponsors and Applicants for PDUFA Products - 3/7/2000

Guidance for Industry: Formal Dispute Resolution: Appeals Above the Division Level - 3/7/2000

International Conference on Harmonsation of Technical Requirements for Registration of Pharmaceuticals for Human Use - 2/10/2000

Guidance for Industry: Possible Dioxin/PCB Contamination of Drug and Biological Products - 8/27/1999

ICH Guidance on Specifications: Test Procedures and Acceptance Criteria for Biotechnological/Biological Products - 8/18/1999

Draft Guidance for Industry: Cooperative Manufacturing Arrangements for Licensed Biologics - 8/3/1999

Draft Guidance for Industry: Interpreting Sameness of Monoclonal Antibody Products Under the Orphan Drug Regulations - 7/24/1999

Guidance for Industry: Container Closure Systems for Packaging Human Drugs and Biologics; Chemistry, Manufacturing, and Controls Documentation - 7/7/1999

Draft Guidance for Industry: INDs for Phase 2 and 3 Studies of Drugs, Including Specified Therapeutic Biotechnology-Derived Products, Chemistry Manufacturing and Controls Content and Format - 4/20/1999

Guidance for Industry: Providing Regulatory Submissions in Electronic Format - General Considerations - 1/28/1999

ICH Guidance on Viral Safety Evaluation of Biotechnology Products Derived From Cell Lines of Human or Animal Origin - 9/24/1998

ICH Guidance on Quality of Biotechnological/Biological Products: Derivation and Characterization of Cell Substrates Used for Production of Biotechnological/Biological Products - 9/21/1998

Guidance for Industry: Environmental Assessment of Human Drug and Biologics Applications - 7/27/1998

Draft Guidance for Industry: Stability Testing of Drug Substances and Drug Products - 6/8/1998

Guidance for Industry: Guidance for Human Somatic Cell Therapy and Gene Therapy - 3/30/1998

Draft Guidance for Industry: Container and Closure Integrity Testing in Lieu of Sterility Testing as a Component of the Stability Protocol for Sterile Products - 1/28/1998

Guidance for Industry for the Submission of Chemistry, Manufacturing, and Controls Information for Synthetic Peptide Substance - 1/16/1998

Guidance for Industry: Q3C Impurities: Residual Solvents - 12/24/1997

Guidance for Industry - The Sourcing and Processing of Gelatin to Reduce the Potential Risk Posed by Bovine Spongiform Encephalopathy (BSE) in FDA-Regulated Products for Human Use - 10/07/1997

Guidance for Industry - Changes to an Approved Application: Biological Products - 7/24/1997

Guidance for Industry - Changes to an Approved Application for Specified Biotechnology and Specified Synthetic Biological Products - 7/24/1997

International Conference on Harmonisation (ICH) Guidelines for the Photostability Testing of New Drug Substances and Products - 5/16/1997

Points to Consider in the Manufacture and Testing of Monoclonal Antibody Products for Human Use - 2/28/1997

Guidance For the Submission of Chemistry, Manufacturing and Controls Information and Establishment Description for Autologous Somatic Cell Therapy Products - 1/10/1997

Points to Consider on Plasmid DNA Vaccines for Preventive Infectious Disease Indications - 12/27/1996

Guidance for Industry for the Submission of Chemistry, Manufacturing, and Controls Information for a Therapeutic Recombinant DNA-Derived Product or a Monoclonal Antibody Product for In Vivo Use - 8/1996

International Conference on Harmonisation: Final Guidance on Stability Testing of Biotechnological / Biological Products - 7/10/1996

FDA Guidance Concerning Demonstration of Comparability of Human Biological Products, Including Therapeutic Biotechnology-Derived Products - 4/1996

International Conference on Harmonisation: Final Guideline on Quality of Biotechnical Products: Analysis of the Expression Construct in Cells Used for the Production of r-DNA Derived Protein Products - 2/1996

Guidance for Industry: Content and Format of Investigational New Drug Applications (INDs) for Phase 1 Studies of Drugs, Including Well-Characterized, Therapeutic, Biotechnology-derived Products - 11/1995

FDA Guidance Document Concerning Use of Pilot Manufacturing Facilities for the Development and Manufacturing of Biological Products - 7/11/1995

Points to Consider in the Manufacture and Testing of Therapeutic Products for Human Use Derived from Transgenic Animals - 1995

Guidance for Industry for the Submission Documentation for Sterilization Process Validation in Applications for Human and Veterinary Drug Products - 11/1994

Points to Consider in the Characterization of Cell Lines Used to Produce Biologicals - 5/17/1993

Supplement to the Points to Consider in the Production and Testing of New Drugs and Biologics Produced by Recombinant DNA Technology: Nucleic Acid Characterization and Genetic Stability - 4/6/1992

Draft Points to Consider in Human Somatic Cell Therapy and Gene Therapy - 8/27/1991

Guideline for the Determination of Residual Moisture in Dried Biological Products - 1/1/1990

Guideline on Validation of the Limulus Amebocyte Lysate Test as an End-Product Endotoxin Test For Human and Animal Parenteral Drugs, Biological Products and Medical Devices - 12/1987

Points to Consider in the Production and Testing of New Drugs and Biologicals Produced by Recombinant DNA Technology - 4/10/1985

Interferon Test Procedures: Points to Consider in the Production and Testing of Interferon Intended for Investigational Use in Humans - 7/28/1983

Table 92. Some CMC regulatory guidances available on the CDER website; www.fda.gov/cder/guidance/index.htm

Chemistry

Changes to an Approved Application for Specified Biotechnology and Specified Synthetic Biological Products 7/1997

Container Closure Systems for Packaging Human Drugs and Biologics 5/1999

Container Closure Systems for Packaging Human Drugs and Biologics -- Questions and Answers Issued 5/2002

Demonstration of Comparability of Human Biological Products, Including Therapeutic Biotechnology-derived Products

Environmental Assessment of Human Drug and Biologics Applications 7/1998

INDs for Phase 2 and Phase 3 Studies Chemistry, Manufacturing, and Controls Information

IND Meetings for Human Drugs and Biologics Chemistry, Manufacturing, and Controls Information 5/2001 Monoclonal Antibodies Used as Reagents in Drug Manufacturing 3/2001

Submission Documentation for Sterilization Process Validation in Applications for Human and Veterinary Drug Products

Submission of Chemistry, Manufacturing, and Controls Information for Synthetic Peptide Substances

Submitting Documentation for the Stability of Human Drugs and Biologics 2/1987

Compliance

Comparability Protocols -- Chemistry, Manufacturing, and Controls Information 2/2003

Stability Testing of Drug Substances and Drug Products 6/5/1998

General Principles of Process Validation

Guideline for Validation of Limulus Amebocyte Lysate Test as an End-Product Endotoxin Test for Human and Animal Parenteral Drugs, Biological Products, and Medical Devices

Sterile Drug Products Produced by Aseptic Processing

Investigating Out of Specification (OOS) Test Results for Pharmaceutical Production 9/30/1998

ICH

Q1A(R) Stability Testing of New Drug Substances and Products

Q1B Photostability Testing of New Drug Substances and Products

Q1C Stability Testing for New Dosage Forms

Q1D Bracketing and Matrixing Designs for Stability Testing of New Drug Substances and Products

Q1E Evaluation of Stability Data

Q1F Stability Data Package for Registration in Climatic Zones III and IV

Q2A Test on Validation of Analytical Procedures

Q2B Validation of Analytical Procedures: Methodology

Q3A Impurities in New Drug Substances

Q3B(R) Impurities in New Drug Products

Q3C Impurities: Residual Solvents

Q5A Viral Safety Evaluation of Biotechnology Products Derived From Cell Lines of Human or Animal Origin

Q5B Quality of Biotechnological Products: Analysis of the Expression Construct in Cells Used for Production of r-DNA Derived Protein Products

Q5C Quality of Biotechnological Products: Stability Testing of Biotechnological/Biological Products

Q5D Quality of Biotechnological Products: Derivation and Characterization of Cell Substrates Used for Production of Biotechnological/Biological Products

Q6A International Conference on Harmonisation; Guidance on Q6A Specifications: Test Procedures and Acceptance Criteria for New Drug Substances and New Drug Products: Chemical Substances.

Q6B Specifications: Test Procedures and Acceptance Criteria for Biotechnological/Biological Products

Q7A Good Manufacturing Practice Guidance for Active Pharmaceutical Ingredients

M4 Common Technical Document--Quality: Questions and Answers/Location Issues

The ORA website contains some FDA 483 reports and provides a number of compliance guides that are resources for the inspectors (Table 93).

Table 93. Inspectional guides available on the ORA website; www.fda.gov/ora/

High Purity Water Systems (7/93)

Lyophilization of Parenterals (7/93)

Microbiological Pharmaceutical Quality Control Labs (7/93)

Pharmaceutical Quality Control Laboratories (7/93)

Validation of Cleaning Processes (7/93)

Biotechnology Inspection Guide

3. RESOURCES FROM EMEA

The EMEA website also provides much biopharmaceutical CMC regulatory compliance resources. The website provides information on the products that have been approved, a list of guidance documents, and safety information (e.g., Warning Letters issued, recalls, Dear Doctor letters, etc.). Just recently, they added a site for the activities of their inspection group. A most useful feature available on the EMEA website is the "What's New" site. Unlike, the FDA websites, there is no automatic email notification of recently added information onto this website. Some of the CMC regulatory guidances available on the EMEA website are presented in Table 93.

Table 94. Some CMC regulatory guidances available on the EMEA website; www.emea.eu.int/index/indexh1.htm

Adopted Guidelines

CPMP/BWP/3207/00 Note for Guidance on Comparability of Medicinal Products containing Biotechnology-derived Proteins as Drug Substance (CPMP adopted Sept. 01)

EMEA/410/01 *Rev. 1* Note for Guidance on Minimising the Risk of Transmitting Animal Spongiform Encephalopathy Agents via Human and Veterinary Medicinal Products, (Adopted by CPMP/CVMP May 2001)

CPMP/BWP/3088/99 Note for Guidance on the Quality, Preclinical and Clinical Aspects of Gene Transfer Medicinal Products (Adopted April 2001)

CPMP/BWP/328/99 Development Pharmaceutics for Biotechnological and Biological Products - Annex to Note for Guidance on Development Pharmaceutics (CPMP/QWP/155/96)

CPMP/BWP/268/95 Note for Guidance on Virus Validation Studies:The Design, Contribution and Interpretation of Studies validating the Inactivation and Removal of Viruses (CPMP adopted Feb.96)

Draft Guidelines

CPMP/3097/02 Note for Guidance *Consultation* on Comparability of Medicinal Products containing Biotechnology-derived Proteins as Drug Substance (Released for consultation July 02)

CPMP/BWP1793/02 Note for Guidance on the Use of Bovine Serum in the manufacture of Human Biological Medicinal Products (Released for Consultation April 02)

Points to Consider

CPMP/BWP/764/02 Points to Consider on Quality aspects on medicinal products containing active substances produced by stable transgene expression in Higher Plants (Adopted, April 2002)

CPMP/BWP/41450/98 Points to Consider on the Manufacture and Quality Control of Human Somatic Cell Therapy Medicinal Products (*Adopted, May 2001*)

4. RESOURCES FROM PROFESSIONAL ASSOCIATIONS

Resources form professional associations should not be overlooked when considering CMC regulatory guidance.

RAPS, in addition to its website news center and offering of numerous courses, also offers an email update service, providing daily email notifications (for a fee) of actions taken by regulatory authorities around the world.

PDA and IABS provide scientific forums and conferences to bring together regulatory personnel, from both the agencies and industry, to discuss major issues impacting the biopharmaceutical industry.

5. CONCLUSION

Biopharmaceutical CMC regulatory compliance will continue to be in flux in the years ahead. It's just part of the nature of the business that we are in. I trust that this book will serve as a useful core resource upon which the future biopharmaceutical CMC regulatory compliance changes can be overlaid.

References

1. U.S. Food and Drug Administration, 1983, *Interferon Test Procedures: Points to Consider in the Production and Testing of Interferon Intended for Investigational Use in Humans*, U.S. FDA, Rockville, MD; www.fda.gov/cber/gdlns/ptcifn.pdf
2. U.S. Food and Drug Administration, 1985, *Points to Consider in the Production and Testing of New Drugs and Biologicals Produced by Recombinant DNA Technology*, U.S. FDA, Rockville, MD; www.fda.gov/cber/gdlns/ptcdna.pdf
3. U.S. Food and Drug Administration, 1991, *Biotechnology Inspection Guide Reference Materials and Training Aids*, U.S. FDA, Rockville, MD; www.fda.gov/ora/inspect_ref/igs/biotech.html
4. U.S. Food and Drug Administration, 1997, *Guidance for Industry: ICH S6 PreclinicalSafety Evaluation of Biotechnology-Derived Pharmaceuticals*, U.S. FDA, Rockville, MD; www.fda.gov/cder/guidance/1859fnl.pdf
5. U.S. Food and Drug Administration, 2002, K. C. Zoon, CBER Update: 100 Years of Regulation; Presentation to Washington Research Group May 2, 2002; U.S. FDA, Rockville, MD; www.fda.gov/cber/summaries/schwab050202.htm
6. U.S. Food and Drug Administration, 2002, *Commemorating 100 Years of Biological Regulation: Science and the Regulation of Biological Products from a Rich History to a Challenging Future*, U.S. FDA, Rockville, MD; www.fda.gov/cber/inside/centennial.htm
7. Ferber, D., 2002, Creeping consensus on SV40 and polio vaccine, *Science* **298**: 725-726.
8. U.S. Food and Drug Administration, 1977, *A Review of Procedures for the Detection of Residual Penicillins in Drugs*, U.S. FDA, Rockville, MD; www.fda.gov/cder/dmpq/penicillin.pdf
9. U.S. Food and Drug Administration, 2000, *Human Gene Therapy: Harsh Lessons, High Hopes*, U.S. FDA, Rockville, MD; www.fda.gov/fdac/features/2000/500_gene.html
10. U.S. Food and Drug Administration, 2003, *FDA Advisory Committee Discusses Steps for Potentially Continuing Certain Gene Therapy Trials That Were Recently Place on Hold*, U.S. FDA, Rockville, MD; www.fda.gov/bbs/topics/answers/2003/ans01202.html
11. U.S. Food and Drug Administration, 2000, *Aventis CropScience Finds Bioengineered Protein in Non-StarLink Corn Seed*, U.S. FDA, Rockville, MD; www.fda.gov/oc/po/firmrecalls/aventis11_00.html

12. U.S. Food and Drug Administration, 2002, *FDA Action on Corn Bioengineered to Produce Pharmaceutical Material*, U.S. FDA Rockville, MD; www.fda.gov/bbs/topics/answers/2002/ans01174.html

13. U.S. Food and Drug Administration, 2003, *FDA Investigates Improper Disposal of Bioengineered Pigs*, U.S. FDA, Rockville, MD; www.fda.gov/bbs/topics/answers/2003/ans01197.html

14. U.S. Food and Drug Administration, 2002, *Therapeutics Office Documents Rapid Product Approvals*, U.S. FDA, Rockville, MD; www.fda.gov/cber/pdufa/therapapprov.htm

15. U.S. Food and Drug Administration, 2002, *Guidance for Industry: Container/ Closure Systems for Packaging Human Drugs and Biologics; Questions and Answers*, U.S. FDA, Rockville, MD; www.fda.gov/cber/gdlns/cntanrq&a.pdf

16. U.S. Food and Drug Administration, 1995, *Guidance for Industry: Content and Format of Investigational New Drug Applications (INDs) for Phase 1 Studies of Drugs, Including Well-Characterized, Therapeutic, Biotechnology-derived Products*, U.S. FDA, Rockville, MD; www.fda.gov/cber/gdlns/ind1.pdf

17. U.S. Food and Drug Administration, 2001, *Guidance for Industry: IND Meetings for Human Drugs and Biologics; Chemistry, Manufacturing, and Controls Information*, U.S. FDA, Rockville, MD; www.fda.gov/cber/gdlns/ind052501.pdf

18. U.S. Food and Drug Administration, 2003, *Draft Guidance for Industry: Comparability Protocols – Chemistry, Manufacturing, and Controls Information*, U.S. FDA, Rockville, MD; www.fda.gov/cber/gdlns/cmprprot.pdf

19. U.S. Food and Drug Administration, 2003, *FDA Completes Final Phase of Planning for Consolidation of Certain Products from CBER to CDER, U.S. FDA*, Rockville, MD; www.fda.gov/bbs/topics/news/2003/new00880.html

20. European Agency for the Evaluation of Medicinal Products, 2001, *CPMP/QWP/848/96 Note for Guidance on Process Validation*, EMEA, London; www.emea.eu.int/pdfs/human/qwp/084896en.pdf

21. European Agency for the Evaluation of Medicinal Products, 2002, *CPMP/EWP/226/02 Concept Paper on the Development of a Committee for Proprietary Medicinal Products (CPMP) Note for Guidance on the Clinical Pharmacokinetic Investigation of the Pharmacokinetics of Peptides and Proteins*, EMEA, London; www.emea.eu.int/pdfs/human/ewp/022602en.pdf

22. U.S. Food and Drug Administration, 2001, *Guidance for Industry: ICH Q7A Good Manufacturing Practice Guidance for Active Pharmaceutical Ingredients*, U.S. FDA, Rockville, MD; www.fda.gov/cder/guidance/4286fnl.pdf

23. U.S. Food and Drug Administration, 2000, *International Conference on Harmonization; Guidance on Q6A Specifications: Test Procedures and Acceptance Criteria for New Drug Substances and New Drug Products: Chemical Substances, Federal Register 65:83041-83063*, U.S. FDA, Rockville, MD; www.fda.gov/ohrms/dockets/98fr/122900d.pdf

24. U.S. Food and Drug Administration, 2003, *Guidance for Industry: Revised Recommendations for the Assessment of Donor Suitability and Blood and Blood Product Safety in Cases of Known or Suspected West Nile Virus Infection*, U.S. FDA, Rockville, MD; www.fda.gov/cber/gdlns/wnvguid.pdf

25. U.S. Food and Drug Administration, 2003, *Guidance for Industry: Recommendations for the Assessment of Donor Suitability and Blood Product Safety in Cases of Suspected Severe Acute Respiratory Syndrome (SARS) or*

Exposure to SARS, U.S. FDA, Rockville, MD;
www.fda.gov/cber/gdlns/sarsbldgd.pdf

26. Copmann, T., Davis, G., Garnick, R., Landis, J., Lubiniecki, A., Massa, T., Murano, G., Seamon, K., and Zezza, D., 2001, One product, one process, one set of specifications: A proven quality paradigm for the safety and efficacy of biologic drugs, *BioPharm* **14(3)**: 14-24.

27. U.S. Food and Drug Administration, 2003, *Improving Innovation in Medical Technology: Beyond 2002*, U.S. FDA, Rockville, MD;
www.fda.gov/bbs/topics/news2003/beyond2002/report.html

28. U.S. Food and Drug Administration, 1996, *FDA Guidance Concerning Demonstration of Comparability of Human Biological Products, Including Therapeutic Biotechnology-Derived Products*, U.S. FDA, Rockville, MD;
www.fda.gov/cber/gdlns/comptest.pdf

29. U. S. Food and Drug Administration, 1997, *Points to Consider in the Manufacture and Testing of Monoclonal Antibody Products for Human Use*, U.S. FDA, Rockville, MD; www.fda.gov/cber/gdlns/ptc_mab.pdf

30. U.S. Food and Drug Administration, 2000, *Guidance for Industry: Formal Meetings with Sponsors and Applicants for PDUFA Products*, U.S. FDA, Rockville, MD;
www.fda.gov/cber/gdlns/mtpdufa.pdf

31. U.S. Food and Drug Administration, 2002, *Manual of Standard Operating Procedures and Policies Regulatory – License Applications: Refusal to File Procedures for Biologic License Applications SOPP 8404*, U.S. FDA, Rockville, MD; www.fda.gov/cber/regsopp/8404.htm

32. European Agency for the Evaluation of Medicinal Products, 2003, *EMEA/H/4260/01 EMEA Guidance for Companies Requesting Scientific Advice (SA) and Protocol Assistance (PA)*, EMEA, London;
www.emea.eu.int/pdfs/human/sciadvice/426001en.pdf

33. U.S. Food and Drug Administration, 2003, *K. C. Zoon, CBER Back to the Future, Presentation to the West Coast Biopharmaceutical Products Meeting, January 8, 2003*, U.S. FDA, Rockville, MD;
www.fda.gov/cber/summaries/010803wcbpkz.pdf

34. U.S. Government Printing Office, 2003, *Title 21 Code of Federal Regulations Part 312.23 Investigational New Drug Application: IND content and format*, U.S. GPO, Washington DC;
www.access.gpo.gov/nara/cfr/waisidx_03/21cfr312_03.html

35. U.S. Food and Drug Administration, 1998, *Guidance for Industry: Guidance for Human Somatic Cell Therapy and Gene Therapy*, U.S. FDA, Rockville, MD;
www.fda.gov/cber/gdlns/somgene.pdf

36. U.S. Food and Drug Administration, 1999, *Draft Guidance for Industry: INDs for Phase 2 and 3 Studies of Drugs, Including Specified Therapeutic Biotechnology-Derived Products; Chemistry, Manufacturing, and Controls Content and Format*, U.S. FDA, Rockville, MD; www.fda.gov/cber/gdlns/indbiodft.pdf

37. U.S. Food and Drug Administration, 1998, *Manual of Policies and Procedures, MAPP 6030.1: IND Process and Review Procedures (Including Clinical Holds)*, U.S. FDA, Rockville, MD; www.fda.gov/cder/mapp/6030-1.pdf

38. U.S. Food and Drug Administration, 2000, *Dear Sponsor of an IND or Master File Using or Producing a Gene Therapy Product*, U.S. FDA, Rockville, MD;
www.fda.gov/cber/ltr/gt030600.pdf

39. U.S. Food and Drug Administration, 2002, *C. J. Joneckis, The Road to Approval: CMC Common Pitfalls, Presentation to BIO 2002 Meeting, June 9-12, 2002*, U.S. FDA, Rockville, MD; www.fda.gov/cber/summaries/bio060902cj.htm

40. U.S. Food and Drug Administration, 1996, *Guidance for Industry: For the Submission of Chemistry, Manufacturing, and Controls Information for a Therapeutic Recombinant DNA-Derived Product or a Monoclonal Antibody Product for In Vivo Use*, U.S. FDA, Rockville, MD; www.fda.gov/cber/gdlns/cmcdna.pdf

41. U.S. Food and Drug Administration, 2001, *Guidance for Industry: ICH M4Q: CTD – Quality*, U.S. FDA, Rockville, MD; www.fda.gov/cder/guidance/4539q.pdf

42. U.S. Food and Drug Administration, 2003, *Draft Guidance for Industry: Drug Product Chemistry, Manufacturing, and Controls Information*, U.S. FDA, Rockville, MD; www.fda.gov/cder/guidance/1215dft.pdf

43. U.S. Food and Drug Administration, 1999, *BLA License Application Approval for Ontak®, Denileukin diftitox: CMC Review*, U.S. FDA, Rockville, MD; www.fda.gov/cber/review/denser020599r2.pdf

44. U.S. Food and Drug Administration, 2002, *FDA 483 Report Inspection of Bayer Corporation, Berkeley, CA, February 26-March 15, 2002*, U.S. FDA, Rockville, MD; obtained under Freedom of Information Act

45. U.S. Government Printing Office, 2003, *Title 21, Code of Federal Regulations: Part 211, Current Good Manufacturing Practice for Finished Pharmaceuticals*, U.S. GPO, Washington DC; www.access.gpo.gov/nara/cfr/waisidx_03/21cfr211_03.html

46. U.S. Food and Drug Administration, 2003, *Food, Drug and Cosmetic (FD&C) Act Chapter 5 – Drugs and Devices: Subchapter A – Drugs and Devices, Section 501 Adulterated Drugs and Devices*, U.S. FDA, Rockville, MD; www.fda.gov/opacom/laws/fdcact/fdcact5a.htm

47. European Commission, 2003, *EudraLex Volume 4 Medicinal Products for Human and Veterinary Use: Good Manufacturing Practices*, EC, Brussels; http://pharmacos.eudra.org/F2/eudralex/vol-4/home.htm

48. European Commission, 2003, *EudraLex Volume 4 Medicinal Products for Human and Veterinary Use: Good Manufacturing Practices Annex 2 Manufacture of Biological Medicinal Products for Human Use*, EC, Brussels; http://pharmacos.eudra.org/F2/eudralex/vol-4/pdfs-en/anx02en.pdf

49. European Commission, 2001, *EudraLex Volume 4 Medicinal Products for Human and Veterinary Use: Good Manufacturing Practices Annex 13 Manufacture of Investigational Medicinal Products*, EC, Brussels; http://pharmacos.eudra.org/F2/eudralex/vol-4/pdfs-en/an13en030303Rev1.pdf

50. U.S. Food and Drug Administration, 2001, *FDA Warning Letter to Bayer Corporation, July 24, 2001*, U.S. FDA, Rockville, MD; www.fda.gov/foi/warning_letters/g1575d.pdf

51. U.S. Food and Drug Administration, 2000, *FDA Warning Letter to Genentech, Inc., December 14, 2000*, U.S. FDA, Rockville, MD; www.fda.gov/foi/warning_letters/m4965n.pdf

52. Neuhaus, J., 2002, Trends in regulatory compliance: EU inspection hot topics, presentation at the PDA Basel 2002 International Congress, PDA, Bethesda, MD

53. U.S. Government Printing Office, 2003, *Title 21, Code of Federal Regulations:*

Part 600.14, Biological Products: Reporting of Biological Product Deviations by Licensed Manufacturers, U.S. GPO, Washington DC; www.access.gpo.gov/nara/cfr/waisidx_03/21cfr600_03.html

54. U.S. Food and Drug Administration, 2001, *Draft Guidance for Industry: Biological Product Deviation Reporting for Licensed Manufacturers of Biological Products Other than Blood and Blood Components*, U.S. FDA, Rockville, MD; www.fda.gov/cber/gdlns/devnbld.pdf

55. U.S. Government Printing Office, 2003, *Title 21, Code of Federal Regulations: Part 314.81, Applications for FDA Approval to Market a new Drug: Other Postmarketing Reports*, U.S. GPO, Washington DC; www.access.gpo.gov/nara/cfr/waisidx_03/21cfr314_03.html

56. U.S. Food and Drug Administration, 2003, *Biological Product Deviation Reports: FY02 Annual Summary*, U.S. FDA, Rockville, MD; www.fda.gov/cber/biodev/bpdrfy02.htm

57. European Agency for the Evaluation of Medicinal Products, 2003, *EMEA/INS/GMP/2260/03 Handling of Reports of Suspected Quality Defects in Medicinal Products for Administration to Humans*, EMEA, London; www.emea.eu.int/Inspections/docs/226003en.pdf

58. U.S. Government Printing Office, 2003, *Title 21, Code of Federal Regulations: Part 312.44, Investigational New Drug Application: Termination*, U.S. GPO, Washington DC; www.access.gpo.gov/nara/cfr/waisidx_03/21cfr312_03.html

59. U.S. Food and Drug Administration, 2001, *FDA Warning Letter to AVAX Technologies, Incorporated, October 30, 2001*, U.S. FDA, Rockville, MD; www.fda.gov/foi/warning_letters/g1925d.pdf

60. U.S. Food and Drug Administration, 2001, *FDA Letter to Cell Therapy Research Foundation, January 11, 2001*, U.S. FDA, Rockville, MD; www.fda.gov/cber/ltr/drlaw011101.htm.

61. European Commission, 2001, *Directive 2001/20/EC Approximation of the Laws, Regulations and Administrative Provisions of the Member States Relating to the Implementation of Good Clinical Practice in the Conduct of Clinical Trials on Medicinal Products for Human Use*, EC, Brussels; http://pharmacos.eudra.org/F2/eudralex/vol-1/new_v1/Dir2001-20_en.pdf

62. U. S. Food and Drug Administration, 2002, *FDA Compliance Program Guidance Manual Program 7356.002, Drug Manufacturing Inspections*, U.S. FDA, Rockville, MD; www.fda.gov/ora/cpgm/7356_002/7356-002final.pdf

63. U.S. Food and Drug Administration, 1998, *CBER Product Approval Information – Licensing Action: Palivizumab Product Review Part I*, U.S. FDA, Rockville, MD; www.fda.gov/cber/review/palimed061998r5a.pdf

64. U.S. Government Printing Office, 2003, *Title 21, Code of Federal Regulations: 21 CFR 610.18(c), General Biological Products Standards: Cell Lines Used for Manufacturing Biological Products*, U.S. GPO, Washington, DC; www.access.gpo.gov/nara/cfr/waisidx_03/21cfr610_03.html

65. U.S. Food and Drug Administration, 1998, *ICH Q5D, Guidance on Quality of Biotechnological/Biological Products: Derivation and Characterization of Cell Substrates Used for Production of Biotechnological/Biological Products*, U.S. FDA, Rockville, MD; www.fda.gov/cder/guidance/92198b.pdf

66. U.S. Food and Drug Administration, 1995, *Points to Consider in the Manufacture*

and Testing of Therapeutic Products for Human Use Derived from Transgenic Animals, U.S. FDA, Rockville, MD; www.fda.gov/cber/gdlns/ptc_tga.txt

67. U.S. Food and Drug Administration, 2002, *Draft Guidance for Industry: Drugs, Biologics, and Medical Devices Derived from Bioengineered Plants for Use in Humans and Animals*, U.S. FDA, Rockville, MD; www.fda.gov/cber/gdlns/bioplant.pdf

68. U.S. Food and Drug Administration, 1996, *ICH Q5B, Quality of Biotechnological Products: Analysis of the Expression Construct in Cells Used for Production of r-DNA Derived Protein Products*, U.S. FDA, Rockville, MD; www.fda.gov/cber/gdlns/ich_rdna.txt

69. U.S. Food and Drug Administration, 1998, *ICH Q5A, Guidance on Viral Safety Evaluation of Biotechnology Products Derived from Cell Lines of Human or Animal Origin*, U.S. FDA, Rockville, MD; www.fda.gov/cber/gdlns/virsafe.pdf

70. European Agency for the Evaluation of Medicinal Products, 2002, *CPMP/BWP/1793/02 Note for Guidance on the Use of Bovine Serum in the Manufacture of Human Biological Medicinal Products*, EMEA, London; www.emea.eu.int/pdfs/human/bwp/179302en.pdf

71. U.S. Food and Drug Administration, 1998, *CBER Product Approval Information – Licensing Action: Infliximab Product Review*, U.S. FDA, Rockville, MD; www.fda.gov/cber/review/inflcen082498r5.pdf

72. Stein. K.E. and Webber, K.O., 2001, The regulation of biologic products derived from bioengineered plants, *Current Opinion in Biotechnology* **12**: 308-311.

73. U.S. Food and Drug Administration, 1991, *CBER Dear Biologic Product Manufacturer*, U.S. FDA, Rockville, MD; www.fda.gov/cber/ltr/bio050391.pdf

74. U.S. Food and Drug Administration, 1996, *Letter to Manufacturers of FDA Regulated Drug/Biological/Device Products*, U.S. FDA, Rockville, MD; www.fda.gov/cber/ltr/fdareg050996.pdf

75. U.S. Food and Drug Administration, 2000, *Letter to Manufacturers of Biological Products*, U.S. FDA, Rockville, MD; www.fda.gov/cber/ltr/bse041900.pdf

76. U.S. Food and Drug Administration, 2003, *Bovine Spongiform Encephalopathy (BSE)*, U.S. FDA, Rockville, MD; www.fda.gov/cber/bse/bse.htm

77. European Agency for the Evaluation of Medicinal Products, 2001, *CPMP/CVMP Note for Guidance on Minimizing the Risk of Transmitting Animal Spongiform Encephalopathy Agents Via Human and Veterinary Medicinal Products*, EMEA, London; www.emea.eu.int/pdfs/vet/regaffair/041001en.pdf

78. European Agency for the Evaluation of Medicinal Products, 2001, *Public Statement on the Evaluation of Bovine Spongiform Encephalopathies (BSE)-Risk Via the Use of Materials of Bovine Origin in or During the Manufacture of Vaccines*, EMEA, London; www.emea.eu.int/pdfs/human/press/pus/047601en.pdf

79. U. S. Food and Drug Administration, 1995, *FDA Guidance Document Concerning Use of Pilot Manufacturing Facilities for the Development and Manufacture of Biological Products*, U.S. FDA, Rockville, MD; www.fda.gov/cber/gdlns/pilot.txt

80. U.S. Food and Drug Administration, 1993, *Points to Consider in the Characterization of Cell Lines Used to Produce Biologics*, U.S. FDA, Rockville, MD; www.fda.gov/cber/gdlns/ptccell.pdf

81. U.S. Food and Drug Administration, 1997, *Establishment Inspection Report of IDEC*

Pharmaceuticals Inspection May 19 – June 3,1997 by CBER, U.S. FDA, Rockville, MD; obtained under Freedom of Information Act

82. U.S. Food and Drug Administration, 2002, *CBER Product Approval Information – Licensing Action: Interferon Beta-1a CMC Product Review Part 2*, U.S. FDA, Rockville, MD; www.fda.gov/cber/review/ifnbser030702r9.pdf

83. Garnick, R.L., 1998, Raw materials as a source of contamination in large-scale cell culture, *Dev. Biol. Stand.* **93**: 21-29

84. U.S. Food and Drug Administration, 1993, *Guide to Inspections: Validation of Cleaning Processes*, U.S. FDA, Rockville, MD; www.fda.gov/ora/inspect_ref/igs/valid.html

85. Hill, L., 2000, Performance qualification of clean-in-place systems: Lessons learned from experience in multiproduct biopharmaceutical manufacturing, *BioPharm* **13(4)**: 24-29

86. United States Pharmacopeia/National Formulary, 2003, *USP 26/NF 21 Official Monographs for Water for Injection and Purified Water*, U.S.P. Convention, Rockville, MD

87. European Directorate for the Quality of Medicine, 2003, *Certification Procedure of the EDQM of the Council of Europe*, European Pharmacopoeia, Strasbourg, France; www.pheur.org/site/page_dynamique.php3?lien=M&lien_page=6&id=1

88. European Commission, 2000, *Amending Directive 94/474/EC Regulating the Use of Material Presenting Risks as Regards Transmissible Spongiform Encephalopathies*, EC, Brussels; http://europa.eu.int/eur-lex/en/com/pdf/2000/en_500PC0378.pdf

89. U.S. Department of Agriculture, 2003, *List of USDA-Recognized Animal Health Status of Countries/Areas Regarding Specific Livestock or Poultry Diseases*, USDA, Riverdale, MD; www.aphis.usda.gov/vs/ncie/country.html

90. U.S. Food and Drug Administration, 1999, *Guidance for Industry: Possible Dioxin/PCB Contamination of Drug and Biological Products*, U.S. FDA, Rockville, MD; http://www.fda.gov/cber/gdlns/dioxpcb.pdf

91. Roscioli, N.A., Renshaw, C.A., Gilbert, A.A., Kerry, C.F. and Probst, P.G., 1999, Environmental monitoring considerations for biological manufacturing, *BioPharm* **12(9)**: 32-40

92. United States Pharmacopeia/National Formulary, 2003, *USP 26/NF 21 General Information Chapter Article 1116 Microbial Evaluation of Clean Rooms and Other Controlled Environments*, U.S.P. Convention, Rockville, MD

93. European Commission, 2003, *EudraLex Volume 4 Medicinal Products for Human and Veterinary Use: Good Manufacturing Practices Annex 1 Revision Manufacture of Sterile Medicinal Products*, EC, Brussels; http://pharmacos.eudra.org/F2/eudralex/vol-4/pdfs-en/revan1vol4.pdf

94. European Agency for the Evaluation of Medicinal Products, 2001, *CPMP/BWP/3088/99 Note for Guidance on the Quality, Preclinical and Clinical Aspects of Gene Transfer Medicinal Products*, EMEA, London; www.emea.eu.int/pdfs/human/bwp/308899en.pdf

95. National Institutes of Health, 2002, *NIH Guidelines for Research Involving Recombinant DNA Molecules, Appendix M: Points to Consider in the Design and Submission of Protocols for the Transfer of Recombinant DNA Molecules into One or More Human Research Participants*, NIH (RAC), Bethesda, MD; www4.od.nih.gov/oba/rac/guidelines_02/NIH_Gdlines_2002prn.pdf

96. U.S. Food and Drug Adminstration, 2003, *FDA Reminder to Scientists Involved in Research with Genetically Engineered Animals*, U.S. FDA, Rockville, MD; www.fda.gov/cvm/index/updates/univletter.htm

97. European Agency for the Evaluation of Medicinal Products, 2002, *CPMP/BWP/764/02 Points to Consider on Quality Aspects of Medicinal Products Containing Active Substances Produced by Stable Transgene Expression in Higher Plants*, EMEA, London; www.emea.eu.int/pdfs/human/bwp/076402en.pdf

98. U.S. Department of Agriculture, 2002, *APHIS Permitting System: Summary of Confinement Measures for Organisms Producing Potential Pharmaceuticals*, USDA, Riverdale, MD; www.aphis.usda.gov/ppq/biotech/pdf/pharma_2000.pdf

99. U.S. Government Printing Office, 2003, *Federal Register Notice for APHIS/USDA: Field Testing of Plants Engineered to Produce Pharmaceutical and Industrial Compounds, March 10, 2003, pages 11337-11340*, U.S. GPO, Washington DC; http://frwebgate1.access.gpo.gov/cgi-bin/waisgate.cgi?WAISdocID=22569720856+5+0+0&WAISaction=retrieve

100. U.S. Food and Drug Administration, 1997, *Establishment Inspection Report of IDECPharmaceuticals Inspection October 6-8, 1997 by CBER*, U.S. FDA, Rockville, MD; obtained under Freedom of Information Act

101. U.S. Food and Drug Administration, 2000, *FDA 483 Report Inspection of Genentech, Inc. February 7-14, 2000*, U.S. FDA, Rockville, MD; obtained under Freedom of Information Act

102. U.S. Food and Drug Administration, 1996, *CBER Summary Basis for Approval: Aldesleukin (Proleukin®)*, U.S. FDA, Rockville, MD; www.fda.gov/cber/sba/aldchir050592S.pdf

103. U.S. Food and Drug Administration, 1999, *ORA Compliance Program Guidance, 7341.001, Inspections of Licensed Therapeutic Drug Products*, U.S. FDA, Rockville, MD; obtained under Freedom of Information Act

104. The Gold Sheet, 2001, Biologic Process Validation Challenges are Discussed by FDA and Industry: Validation of Virus Removal (K. Webber), **35(1)**: 25-30; F-D-C Reports, Inc., Chevy Chase, MD

105. O'Leary, R.M., Feuerhelm, D., Peers, D., Xu, Y., and Blank, G.S., 2001, Determining the useful lifetime of chromatography resins: Preparative small-scale studies, *BioPharm* **14(9)**: 10-18.

106. U.S. Government Printing Office, 2003, *Title 21, Code of Federal Regulations: 21 CFR 610.15(b), General Biological Products Standards: Constituent Materials*, U.S. GPO, Washington, DC; www.access.gpo.gov/nara/cfr/waisidx_03/21cfr610_03.html

107. U. S. Food and Drug Administration, 2001, *Guidance for Industry: Monoclonal Antibodies Used as Reagents in Drug Manufacturing*, U.S. FDA, Rockville, MD; www.fda.gov/cber/gdlns/mab032901.pdf

108. European Agency for the Evaluation of Medicinal Products, 1997, *CPMP/BWP/382/97 CPMP Position Statement on DNA and Host Cell Proteins (HCP) Impurities, Routine Testing Versus Validation Studies*, EMEA, London; www.emea.eu.int/pdfs/human/press/pos/038297en.pdf

109. U.S. Food and Drug Administration, 1999, *CBER Summary Basis for Approval: Coagulation Factor VIIa (Recombinant) (rFVIIa) (NovoSeven®)*, U.S. FDA, Rockville, MD; www.fda.gov/cber/sba/viianov032599S.pdf

110. Sofer, G., Lister, D.C., and Boose, J.A., 2003, Virus inactivation in the 1990s – and into the 21st century, Part 6, Inactivation methods grouped by virus, *BioPharm* **16(4)**: 42-52.

111. U.S. Food and Drug Administration, 2002, *Guidance for Industry: Recommendations for Deferral of Donors and Quarantine and Retrieval of Blood and Blood Products in Recent Recipients of Smallpox Vaccine (Vaccinia Virus) and Certain Contacts of Smallpox Vaccine Recipients*, U.S. FDA, Rockville, MD; www.fda.gov/cber/gdlns/smpoxdefquar.pdf

112. U.S. Food and Drug Administration, 2003, *Monkeypox Virus Infections and Blood and Plasma Donors,* U.S. FDA, Rockville, MD; www.fda.gov/cber/infosheets/monkeypox.htm

113. U.S. Food and Drug Administration, 1998, *FDA Establishment Inspection Report, Inspection of MedImmune, Inc., March 26-April 6, 1998*, U.S. FDA, Rockville, MD; obtained under Freedom of Information Act

114. U.S. Food and Drug Administration, 1999, *FDA Warning Letter to IDEC Pharmaceuticals, November 8, 1999*, U.S. FDA, Rockville, MD; www.fda.gov/foi/warning_letters/m3197n.pdf

115. U.S. Food and Drug Administration, 2000, *FDA Establishment Inspection Report, Inspection of Genentech, Inc., August 7-24, 2000*; U.S. FDA, Rockville, MD; obtained under Freedom of Information Act

116. U.S. Government Printing Office, 2003, *Title 21, Code of Federal Regulations: 21 CFR 210.3 Current Good Manufacturing Practice in Manufacturing, Processing, Packing or Holding of Drugs: General -- Definitions*, U.S. GPO, Washington, DC; www.access.gpo.gov/nara/cfr/waisidx_03/21cfr210_03.html

117. U. S. Food and Drug Administration, 2002, *FDA Import Alert #55-02: Increased Surveillance of Glycerin Due to the Presence of Diethylene Glycol*, U. S. FDA, Rockville, MD; www.fda.gov/ora/fiars/ora_import_ia5502.html

118. U.S. Food and Drug Administration, 2002, *Draft Guidance for Industry: Nonclinical Studies for Development of Pharmaceutical Industries*, U.S. FDA, Rockville, MD; www.fda.gov/cber/gdlns/dvpexcp.pdf

119. U.S. Food and Drug Administration, 2003, *CDER Ingredient Search for Approved Drug Products*, U.S. FDA, Rockville, MD; www.accessdata.fda.gov/scripts/cder/iig/index.cfm

120. U.S. Food and Drug Administration, 1997, *BLA License Application Approval for Rituxan®, Rituximab: Product Review*, U.S. FDA, Rockville, MD; www.fda.gov/cber/review/ritugen112697-r2.pdf

121. U.S. Food and Drug Administration, 2003, *Guidance for Industry: Sterile Drug Products Produced by Aseptic Processing (Draft)*, U.S. FDA, Rockville, MD; www.fda.gov/cder/guidance/1874dft.pdf

122. U.S. Food and Drug Administration, 1987, *Guideline on Sterile Drug Products Produced by Aseptic Processing*, U.S. FDA, Rockville, MD; www.fda.gov/cder/guidance/old027fn.pdf

123. U.S. Food and Drug Administration, 1994, *Guidance for Industry: For the Submission Documentation for Sterilization Process Validation in Applications for Human and Veterinary Drug Products*, U.S. FDA, Rockville, MD; www.fda.gov/cder/guidance/cmc2.pdf

124. Eudy, J., 2003, Clean manufacturing cleanrooms: Human contamination, *Adv. Appl. Contamination Control* **6(4)**: 7-11

125. U.S. Food and Drug Administration, 1998, *Warning Letter to Centocor, Inc., July 10, 1998*, U.S. FDA, Rockville, MD; www.fda.gov/foi/warning_letters/d1924b.pdf

126. U.S. Food and Drug Administration, 1999, *Guidance for Industry ICH Q6B Specifications: Test Procedures and Acceptance Criteria for Biotechnological/Biological Products*, U.S. FDA, Rockville, MD; www.fda.gov/cder/guidance/Q6Bfnl.PDF

127. U.S. Government Printing Office, 2003, *Title 21, Code of Federal Regulations: 21 CFR 600.3, Biological Products: General*, U.S. GPO, Washington, DC; www.access.gpo.gov/nara/cfr/waisidx_03/21cfr600_03.html

128. Schenerman, M.A., Hope, J.N., Kletke, C., Singh, J.K., Kimura, R., Tsao, E.I. and Foiena-Wasserman, G., 1999, Comparability testing of a humanized monoclonal antibody (Synagis®) to support cell line stability, process validation, and scale-up for manufacturing, *Biologicals* **27**: 203-215

129. Ma, S. and Nashabeh, W., 1999, Carbohydrate analysis of a chimeric recombinant monoclonal antibody by capillary electrophoresis with laser-induced fluorescence detection, *Anal. Chem.* **71**: 5185-5192.

130. Reed, H., 2003, Analytical characterization of monoclonal antibodies: Linking structure to function, presentation at the FDA/IABS/NIBSC Workshop on State of the Art Analytical Methods for the Characterization of Biological Product Assessment of Comparability June 10-13, 2003, U.S. FDA, Rockville, MD

131. U.S. Government Printing Office, 2003, *Title 21, Code of Federal Regulations: 21 CFR 211.160: Laboratory Controls; General Requirements*, U.S. GPO, Washington, DC; www.access.gpo.gov/nara/cfr/waisidx_03/21cfr211_03.html

132. U.S. Government Printing Office, 2003, *Title 21, Code of Federal Regulations: 21 CFR 610, General Biological Products Standards*, U.S. GPO, Washington, DC www.access.gpo.gov/nara/cfr/waisidx_03/21cfr610_03.html

133. U.S. Government Printing Office, 2003, *Title 21, Code of Federal Regulations: 21 CFR 610.11, General Biological Products Standards; General Safety*, U.S. GPO, Washington, DC; www.access.gpo.gov/nara/cfr/waisidx_03/21cfr610_03.html

134. U.S. Government Printing Office, 2003, *Title 21, Code of Federal Regulations: 21 CFR 601.2(c)(3), Licensing: Applications for Biologics Licenses; Procedures for Filing*, U.S. GPO, Washington, DC; www.access.gpo.gov/nara/cfr/waisidx_03/21cfr601_03.html

135. European Agency for the Evaluation of Medicinal Products, 2002, *European Public Assessment Report (EPAR): Rebif, Serano, Scientific Discussion*, EMEA, London; www.eudra.org/humandocs/Humans/EPAR/Rebif/Rebif.htm

136. European Agency for the Evaluation of Medicinal Products, 2002, *European Public Assessment Report (EPAR): Zenapax,Roche, Scientific Discussion*, EMEA, London; www.eudra.org/humandocs/Humans/EPAR/Zenapax/Zenapax.htm

137. U.S. Government Printing Office, 2002, *Title 21, Code of Federal Regulations: 21 CFR 610.10, General Biological Products Standards; Potency*, U.S. GPO, Washington, DC; www.access.gpo.gov/nara/cfr/waisidx_03/21cfr610_03.html

138. Gazzano-Santoro, H., Ralph, P., Ryskamp, T.C., Chen, A.B., and Mukku, V.R., 1997, A non-radioactive complement-dependent cytotoxicity assay for anti-CD20 monoclonal antibody, *J. Immunol.. Methods,* **202**: 163-171.

139. California Separation Science Society, 2003, *West Coast Biopharmaceutical*

Products (WCBP) CMC Strategy Forum #2, January 6, 2003: Summary of the Afternoon Discussion on Potency, CASSS, South San Francisco, CA; www.casss.org/discussion/cmc_pdf/FinalDraftCMCReview010603PotencyforPu blication.pdf

140. United States Pharmacopeia/National Formulary, 2003, *USP 26/NF 21 Official Monographs for Human Insulin and Human Insulin Injection*, U.S.P. Convention, Rockville, MD

141. U.S. Food and Drug Administration, 1995, *Guideline for Industry: ICH Q2A Text on Validation of Analytical Procedures*, U.S. FDA, Rockville, MD; www.fda.gov/cder/guidance/ichq2a.pdf

142. U.S. Food and Drug Administration, 1996, *Guidance for Industry: ICH Q2B Validation of Procedures: Methodology*, U.S. FDA, Rockville, MD; www.fda.gov/cder/guidance/1320fnl.pdf

143. U.S. Food and Drug Administration, 2000, *Draft Guidance for Industry: Analytical Procedures and Methods Validation: Chemistry, Manufacturing, and Controls Documentation*, U.S. FDA, Rockville, MD; www.fda.gov/cder/guidance/2396dft.pdf

144. United States Pharmacopeia/National Formulary, 2003, *USP 26/NF 21 Article 1225: Validation of Compendial Methods*, U.S.P. Convention, Rockville, MD

145. Bongers, J., Cummings, J.J., Ebert, M.B., Federici, M.M., Gledhill, L., Gulati, D., Hilliard, G.M., Jones, B.H., Lee, K.R., Mozdzanowski, J., Naimoli, M. and Burman, S., 2000, Validation of a peptide mapping method for a therapeutic monoclonal antibody: What could we possibly learn about a method we have run 100 times?, *J. Pharm. Biomed. Anal.* **21**: 1099-1128

146. Allen, D., Baffi, R., Bausch, J., Bongers, J., Costello, M., Dougherty, J., Jr., Federici, M., Garnick, R., Peterson, S., Riggins, R., Sewerin, K., and Tuls, J., 1996, Validation of peptide mapping for protein identity and genetic stability, *Biologicals* **24**: 255-275

147. Kotts, C.E., Gilkerson, E., Trinh Jr., D., Hawker, G., Chen, A. and Gazzano Santoro, H., 1998, Assay validation report: Cell proliferation test for human growth hormone, *Pharmacopeial Forum* **25**: 8313-8332

148. United States Pharmacopeia/National Formulary, 2003, *USP 26/NF 21 Article 1047: Biotechnology-Derived Articles – Tests*, U.S.P. Convention, Rockville, MD

149. U.S. Food and Drug Administration, 1993, *FDA Guide to Inspections of Pharmaceutical Quality Control Laboratories*, U.S. FDA, Rockville, MD; www.fda.gov/ora/inspect_ref/igs/pharm.html

150. U.S. Food and Drug Administration, 1993, *FDA Guide to Inspections of Microbiological Pharmaceutical Quality Control Laboratories*, U.S. FDA, Rockville, MD; www.fda.gov/ora/inspect_ref/igs/micro.html

151. Ritter, N., and Wiebe, M., 2001, Validating critical reagents used in CGMP analytical testing: Ensuring method integrity and reliable assay performance, *BioPharm* **14(5)**: 12-21.

152. Williamsburg BioProcessing Foundation, Reference Material Project: Adenoviral Reference Material – Characterization, WilBio, Virginia Beach, VA; www.wilbio.com/

153. National Institute of Biological Standards and Control, Catalogue of International

Biological Standards, NIBSC, South Mimms, United Kingdom;
www.nibsc.ac.uk/catalog/standards

154. Williamsburg BioProcessing Foundation, Reference Material Project: Adenoviral
Reference Material (ATCC VR-1516), WilBio, Virginia Beach, VA;
www.wilbio.com/

155. U.S. Government Printing Office, 2003, *Title 21, Code of Federal Regulations: 21
CFR 211.160(b), Laboratory Controls*, U.S. GPO, Washington, DC;
www.access.gpo.gov/nara/cfr/waisidx_03/21cfr211_03.html

156. U.S. Government Printing Office, 2003, *Title 21, Code of Federal Regulations: 21
CFR 211.180(e), Records and Reports: General Requirements*, U.S. GPO,
Washington, DC; www.access.gpo.gov/nara/cfr/waisidx_03/21cfr211_03.html

157. U.S. Food and Drug Administration, 2003, *Guidance for Industry: ICH Q3A
Impurities in New Drug Substances*, U.S. FDA, Rockville, MD;
www.fda.gov/cder/guidance/4164fnl.pdf

158. U.S. Food and Drug Adminstration, 2001, *FDA Letter to Sponsors Using Vero
Cells as a Cell Substrate for Investigational Vaccines*, U.S. FDA, Rockville,
MD; www.fda.gov/cber/ltr/vero031301.pdf

159. Krause, P.R., and Lewis, A.M., Jr., 1998, Safety of viral DNA in biological
products, *Biologicals* **26**: 317-320.

160. U.S. Government Printing Office, 2003, *Title 21, Code of Federal Regulations:
21CFR 610.15(b), General Biological Product Standards: Constituent
Materials*, U.S. GPO, Washington, DC;
www.access.gpo.gov/nara/cfr/waisidx_03/21cfr610_03.html

161. U.S. Food and Drug Administration, 1987, *Guideline on Validation of the Limulus
Amebocyte Lysate Test as an End-Product Endotoxin Test for Human and
Animal Parenteral Drugs and Biological Products and Medical Devices*, U.S.
FDA, Rockville, MD; www.fda.gov/cber/gdlns/lal.pdf

162. U.S. Centers for Disease Control and Prevention, 1998, *Morbidity and Mortality
Weekly Report, July 31, 1998, Clinical Sepsis and Death in a Newborn Nursery
Assoicatied with Contaminated Parenteral Medications – Brazil, 1996*, U.S.
CDC, Atlanta, GA; www.cdc.gov/mmwr/preview/mmwrhtml/00054058.htm

163. U.S. Government Printing Office, 2003, *Title 21, Code of Federal Regulations:
21CFR 610.13, General Biological Product Standards: Purity*, U.S. GPO,
Washington, DC; www.access.gpo.gov/nara/cfr/waisidx_03/21cfr610_03.html

164. U.S. Food and Drug Administration, 1990, *Guideline for the Determination of
Residual Moisture in Dried Biological Products*, U.S. FDA, Rockville, MD;
www.fda.gov/cber/gdlns/moisture.pdf

165. U.S. Government Printing Office, 2003, *Title 21, Code of Federal Regulations:
21CFR 610.61(m), General Biological Product Standards: Package Label* , U.S.
GPO, Washington, DC;
www.access.gpo.gov/nara/cfr/waisidx_03/21cfr610_03.html

166. Geigert, J., 1997, Appropriate specifications at the IND stage, *PDA J. Pharm. Sci.
& Technol.*, **51**: 78-80

167. U.S. Food and Drug Administration, 2001, *CBER Product Approval Letter to Eli
Lilly and Company for Xigris®*, U.S. FDA, Rockville, MD;
www.fda.gov/cber/approvltr/droteli112101L.pdf

168. U.S. Food and Drug Administration, 2001, *CBER Product Approval*

Letter to Amgen for Aranesp^{TM}, U.S. FDA, Rockville, MD;
www.fda.gov/cber/approvltr/darbamg0917011.pdf

169. U.S. Food and Drug Administration, 2001, *CBER BLA License Application Approval for for Aranesp^{TM} : CMC Review,* U.S. FDA, Rockville, MD;
www.fda.gov/cber/review/darbamg091701r5.pdf

170. U.S. Food and Drug Administration, 2002, *CBER Supplement Product Approval Letter to Amgen for Aranesp^{TM}*, U.S. FDA, Rockville, MD;
www.fda.gov/cber/approvltr/darbamg071902L.pdf

171. U.S. Food and Drug Administration, 2003, *CBER Product Approval Letter to Biomarin Pharmceutical for Aldurazyme®*, U.S. FDA, Rockville, MD;
www.fda.gov/cber/approvltr/larobio043003L.htm

172. U.S. Food and Drug Administration, 1996, *Guideline for Industry: Quality of Biotechnological Products: Stability Testing of Biotechnological/Biological Products*, U.S. FDA, Rockville, MD; www.fda.gov/cder/guidance/ichq5c.pdf

173. U.S. Food and Drug Administration, 2001, *Guidance for Industry: ICH Q1A Stability Testing of New Drug Substances and Products*, U.S. FDA, Rockville, MD; www.fda.gov/cder/guidance/4282fnl.pdf

174. U.S. Food and Drug Administration, 1997, *BLA License Application Approval for Becaplermin Concentrate: CMC Review*, U.S. FDA, Rockville, MD;
www.fda.gov/cber/review/becachi121697-r1.pdf

175. U.S. Government Printing Office, 2003, *Title 21, Code of Federal Regulations: 21CFR 211.137, Expiration Dating*, U.S. GPO, Washington, DC;
www.access.gpo.gov/nara/cfr/waisidx_03/21cfr211_03.html

176. U.S. Food and Drug Administration, 2002, *Draft ICH Q1E Evaluation of Stability Data*, U.S. FDA, Rockville, MD; www.fda.gov/cder/guidance/4983dft.pdf

177. The Gold Sheet, 2003, Drug Quality Initiative on Pace at Half-Year Mark: Laboratory System Remains Most Commonly Cited, **37(4)**: 15-17; F-D-C Reports, Inc., Chevy Chase, MD

178. U.S. Food and Drug Administration, 1998, *Draft Guidance for Industry: Investigating Out of Specification (OOS) Test Results for Pharmaceutical Production*, U.S. FDA, Rockville, MD; www.fda.gov/cder/guidance/1212dft.pdf

179. U.S. Food and Drug Administration, 1999, *FDA Warning Letter to Genetics Institute, December 23, 1999*, U.S. FDA, Rockville, MD;
www.fda.gov/foi/warning_letters/m3321n.pdf

180. U.S. Food and Drug Administration, 2001, *FDA 483 Report Inspection of Genentech, Inc., July 24 – August 3, 2001*, U.S. FDA, Rockville, MD; obtained under Freedom of Information Act

181. U.S. Food and Drug Administration, 2001, *FDA 483 Report Inspection of Eli Lilly and Company, October 15 – November 14, 2001*, U.S. FDA, Rockville, MD; obtained under Freedom of Information Act

182. U.S. Food and Drug Administration, 2003, *FDA Warning Letter to Jacqueline M. Halton, M.D., Children's Hospital of Eastern Ontario, April 14, 2003*, U.S. FDA, Rockville, MD; www.fda.gov/foi/warning_letters/g3946d.pdf

183. U.S. Food and Drug Administration, 2002, *Therapeutics Office Product Approvals: Reasons for Multiple Review Cycles,* U.S. FDA, Rockville, MD;
www.fda.gov/cber/pdufa/revcycles.htm

184. U.S. Government Printing Office, 2003, *Title 21, Code of Federal Regulations:*

21CFR 601.12, Licensing: Changes to an Approved Application, U.S. GPO, Washington, DC; www.access.gpo.gov/nara/cfr/waisidx_03/21cfr601_03.html

185. U.S. Government Printing Office, 2003, *Title 21, Code of Federal Regulations: 21CFR 314.70(g): Applications for FDA Approval to Market a New Drug: Supplements and Other Changes to an Approved Application*, U.S. GPO, Washington, DC; www.access.gpo.gov/nara/cfr/waisidx_03/21cfr314_03.html

186. U.S. Food and Drug Administration, 1997, *Guidance for Industry: Changes to an Approved Application for Specified Biotechnology and Specified Synthetic Biological Products*, U.S. FDA, Rockville, MD; www.fda.gov/cber/gdlns/chbiosyn.pdf

187. European Commission, 2003, *EudraLex Volume 2A Notice to Applicants – Procedures for Marketing Authorization: Chapter 5 Variations*, EC, Brussels; http://pharmacos.eudra.org/F2/eudralex/vol-2/A/pdfs-en/cap5aen.pdf

188. European Commission, 2003, *EudraLex Volume 2C Notice to Applicants – Regulatory Guidelines: Guideline on Dossier Requirements for Type I Variations*, EC, Brussels; http://pharmacos.eudra.org/F2/eudralex/vol-2/C/NTAnov99.pdf

189. European Commission, 2003, *Commission Regulation (EC) No 1085/2003 of 3 June 2003: New Variation Regulation*, EC, Brussels; http://pharmacos.eudra.org/F2/review/doc/2003_June/direct_comm_2003_63_en%20.pdf

190. European Agency for the Evaluation of Medicinal Products, 2003, *European Public Assessment Report (EPAR): Betaferon, Schering AG, Steps Taken After Authorization*, EMEA, London; www.eudra.org/humandocs/Humans/EPAR/Betaferon/Betaferon.htm

191. European Agency for the Evaluation of Medicinal Products, 2001, *European Public Assessment Report (EPAR): Mabthera, Roche, Steps Taken After Authorization*, EMEA, London; www.eudra.org/humandocs/Humans/EPAR/Mabthera/Mabthera.htm

192. European Agency for the Evaluation of Medicinal Products, 2001, *CPMP Note for Guidance on Comparability of Medicinal Products Containing Biotechnology-Derived Proteins as Drug Substance*, EMEA, London; www.emea.eu.int/pdfs/human/bwp/320700en.pdf

193. European Agency for the Evaluation of Medicinal Products, 2002, *CPMP Note for Guidance on Comparability of Medicinal Products Containing Biotechnology-Derived Proteins as Drug Substance: Annex on Non-Clinical and Clinical Considerations*, EMEA, London; www.emea.eu.int/pdfs/human/ewp/309702en.pdf

194. U.S. Food and Drug Administration, 2002, *CBER Product Approval Letter to Abbott Laboratories for Adalimumab (Humira^{TM})*, Rockville, MD; www.fda.gov/cber/approvltr/adalabb123102L.htm

195. The Gold Sheet, 1999, Final 314.70 Companion Guidance Targeted for November: Comparability Protocols Build Into New Regulations, **33(10)**: 9-14; F-D-C Reports, Inc., Chevy Chase, MD

196. Vanderlaan, M., Keck, R., Harris, R., and Stark D., 2003, Case studies in comparability, presentation at the PDA Spring Conference in San Diego March 17-21, 2003, PDA, Bethesda, MD

197. Kuhne, W., 2001, Site transfer and process validation in case of Herceptin active

pharmaceutical ingredient (API), presentation at the PDA/IABS Conference in Berlin September 6-7, 2001, PDA, Bethesda, MD

198. Waters, D., 2003, Manufacturing process changes and product comparability: A bulk drug scale-up PAS requiring a pre-approval inspection, presentation at the PDA Spring Conference in San Diego March 17-21, 2003, PDA Bethesda, MD

199. European Agency for the Evaluation of Medicinal Products, 2003, *European Public Assessment Report (EPAR): Xigris Eli Lilly and Company, Scientific Discussion*, EMEA, London; www.eudra.org/humandocs/Humans/EPAR/xigris/xigris.htm

200. U.S. Food and Drug Administration, 1996, *CBER Summary Basis of Approval for Interferon beta-1a (Avonex®)*, U.S. FDA, Rockville, MD; www.fda.gov/cber/sba/ifnbbio051796s.pdf

201. U.S. Food and Drug Administration, 2002, *CBER Product Approval Information for Interferon beta-1a (Avonex®): Label*, U.S. FDA, Rockville, MD; www.fda.gov/cber/label/ifnbbio111202LB.pdf

202. Mire-Sluis, A.R., 2003, Current perspectives on the use of analytical data to show comparability, presentation at the FDA/IABS/NIBSC Workshop on State of the Art Analytical Methods for the Characterization of Biological Product Assessment of Comparability June 10-13, 2003, U.S. FDA, Rockville, MD

203. Kelley, G., 2002, EPO saga to augur regulatory change?, *BioPharm International* **15(12)**: 44

204. Shacter, E., 2003, Regulatory perspectives on product comparability: Lessons from changes in raw materials, presentation at the PDA Spring Conference in San Diego March 17-21, 2003, PDA Bethesda, MD

205. Siegel, J.P., 2002, Comparability of biotechnology derived protein products: Lessons from the U.S. experience, presentation at the Drug Information Association (DIA) Basel Conference, DIA, Horsham, PA

206. Roth, G.Y., 2003, Biomanufacturing report: What's going on in the bio-outsourcing industry?, *Contract Pharma* June 2003, 55-58

207. U.S. Government Printing Office, 2003, *Title 21, Code of Federal Regulations: 21CFR 600.3(t) Biological Products: General*, U.S. GPO, Washington, DC; www.access.gpo.gov/nara/cfr/waisidx_03/21cfr600_03.html

208. U.S. Government Printing Office, 2003, *Title 21, Code of Federal Regulations: 21CFR 211.22(a) Organization and Personnel: Responsibilities of Quality Control Unit:*, U.S. GPO, Washington, DC; www.access.gpo.gov/nara/cfr/waisidx_03/21cfr211_03.html

209. U.S. Food and Drug Administration, 1999, *Draft Guidance for Industry: Cooperative Manufacturing Arrangements for Licensed Biologics*, U.S. FDA, Rockville, MD; www.fda.gov/cber/gdlns/coopmfr.pdf

210. European Commission, 2003, *EudraLex Volume 4 Medicinal Products for Human and Veterinary Use: Good Manufacturing Practices Chapter 7 Contract Manufacture and Analysis*, EC, Brussels; http://pharmacos.eudra.org/F2/eudralex/vol-4/pdfs-en/cap7en.pdf

211. Bobrowicz, G., 2001, The quality agreement: Compliance considerations in selecting a contract manufacturer, *BioPharm* **14(2)**: 14-20.

212. The Gold Sheet, 2002, Early Drafting of Separate Quality Agreements with Contractors Urged, **36(11)**: 1-20; F-D-C Reports, Inc., Chevy Chase, MD

213. Roth, G.Y., 2002, Ask the Board: Where does QA rank?, *Contract Pharma* March

2002, 26-28

214. U.S. Food and Drug Administration, 2003, FDA Summary Progress Report
     Pharmaceutical cGMPs for the 21st Century: A Risk-Based Approach, U.S.
     FDA, Rockville, MD; www.fda.gov/cder/gmp/21stcenturysummary.htm

215. U.S. Food and Drug Administration, 2003, *Guidance for Industry: Comparability
     Protocols – Protein Drug Products and Biological Products – Chemistry,
     Manufacturing, and Controls Information,* U.S. FDA, Rockville, MD;
     www.fda.gov/cber/gdlns/protcmc.pdf

Index